Library of Ethics and Applied Philosophy

Volume 42

The Library of Ethics and Applied Philosophy addresses a broad range of topical issues emerging in practical philosophy, such as ethics, social philosophy, political philosophy, philosophy of action, It focuses on the role of scientific research and emerging technologies, and combines case studies with conceptual and methodological debates. We are facing a global crisis, which raises a plethora of normative issues, but also poses a challenge to existing conceptual and methodological resources of academic philosophy. The series aims to contribute to a philosophical diagnostic of the present by exploring the impact of emerging techno-scientific developments on zeitgeist, collective self-image, and worldview. Concrete and urgent emerging issues will be addressed in depth and from a continental philosophical perspective, which may include dialectics,hermeneutics, phenomenology, post-phenomenology, psychoanalysis, critical theory and similar approaches.

Nicholas Agar • Stuart Whatley • Dan Weijers

How to Think about Progress

A Skeptic's Guide to Technology

 Springer

Nicholas Agar
Hamilton, New Zealand

Stuart Whatley
Cary, NC, USA

Dan Weijers
Hamilton, New Zealand

ISSN 1387-6678   ISSN 2215-0323   (electronic)
Library of Ethics and Applied Philosophy
ISBN 978-3-031-68937-6   ISBN 978-3-031-68938-3   (eBook)
https://doi.org/10.1007/978-3-031-68938-3

This Springer imprint is published by the registered company Springer Nature Switzerland AG
The registered company address is: Gewerbestrasse 11, 6330 Cham, Switzerland

If disposing of this product, please recycle the paper.

Preface

This book presents a way to think about the future—and to prepare for it. We need more intellectually serious advice about forthcoming challenges, and that requires a fresh generation who can think wisely about a wide range of issues unknown to earlier generations. In particular, this book criticizes an overly optimistic view of our ability to progress technology quickly and solve these problems "any day now." Examples of overly optimistic predictions of progress in important domains are presented and shown to have failed over and over. We should think about progress carefully, learning from the past and understanding why we find optimistic exhortations about technological fixes to complex problems so appealing. To achieve safe and sustainable progress, we should consider individual, social, and political solutions alongside technological ones.

Thinking appropriately about progress is something we should all engage in. Engaging in an informed way is not always easy. Sound bites and headlines are understandable but too simplified. All too often they also echo or even exaggerate the claims of technological optimists, many of which stand to gain financially from optimism about the technological solution to a major problem they are working on. Gaining detailed information directly from a range of (hopefully less biased) experts is also difficult for most people. The unfriendly conventions of academic journals have created a walled garden that admits entry only to a select few. Perhaps what we need more of is deep but accessible work that draws together expert research and applies it to our significant current and future troubles.

This book charts a middle path between the immediacy of communications in the internet age and the mannered prose of the academy. It aims for an intellectually serious but accessible form of engaging with the future. When claims are made, evidence must be cited. If you want to assert that levels of subjective well-being are inversely correlated with time spent on Facebook, you should point to studies supporting that claim. But excessive citations as a means of special pleading go beyond this standard. The obligation to offer evidence need not conform to the baroque conventions and stylistic disfigurements of academic writing.

We offer a model of writing about the future that lowers the barriers to entry for fresh thinking. When people venture ideas about an essentially uncertain future, they should be aware that they risk being wrong. But this makes them no different from many of the self-appointed experts of our age. There should come a time when we can render an accounting of our forecasts, failed and successful. We will eventually know whether Elon Musk succeeded in creating a city of one million people on Mars by 2061, as he has said he will do. Suppose we arrive at 2100 and find Mars free of human settlers. Musk's forecast will have failed. But that would not gainsay the value of his forecast. There may still be value in his statements about the future, even if they fail as forecasts, because they offer speculation about humanity's potential, and such conjectures sometimes shape material realities (indeed, Musk himself was originally inspired by the sci-fi author Isaac Asimov's Foundation series). The main complaint we have is not that Musk offers such a daring vision of our future. It is that his exciting visions of the future tend to be privileged, simply owing to his celebrity status.

We need the contrasting and competing visions of the many, including young minds that enter universities with an eagerness to solve our species' problems. Musk's billions embolden him to venture forecasts about the future without fear of being falsified. People who lack his means but share his desire to describe and shape humanity's future should be similarly empowered. We need more, not fewer, ideas about an essentially uncertain future. It would be both tragic and dangerous if we failed to see an opportunity that was within the considerable imaginative range of our species, simply because we feared being wrong.

What Generative AI (Artificial Intelligence) Means for Thinking and Writing

This ethos inspired our writing of the book. But we, along with many thinkers, were taken entirely off guard by the arrival of generative AIs such as ChatGPT, which have generated feverish handwringing about the future of work, education, and much else. ChatGPT has indeed introduced a novel anxiety about written language. It produces text that seems eloquently expressed and eerily human. Following its arrival, there was no shortage of commentaries in the mainstream media that included a confession ("The text above was written by ChatGPT") or a disclaimer ("No, this was not written by ChatGPT").

ChatGPT also produces work that bears many of the hallmarks of good scholarship. Students are already submitting ChatGPT's output in their courses, and there are now ChatGPT contributions to the academic literature. Many scholars, panicked by the arrival of generative AI, may shift their focus to what it might do next. If generative AI can write good student essays today, it might pen the solutions to all of our problems tomorrow and be excellent at explaining them. This would leave very little for academics to do in their professional lives. So, focusing on what generative AI might do next seems rational. If you find yourself in a paddock about to be charged by a bull, you will focus on what the animal will do next. You won't be worrying about how the bull's descendants might respond to your own progeny.

But our interest in how people think and whether they generate their ideas for themselves does look to the future, beyond today's tenured professors and the fears they may harbor about generative AI. As more and more of the predictions about progress and professed solutions to the world's problems are generated by AI, the average person will be even more inundated and overwhelmed with information than ever before. AI might offer to summarize this information for us, but in doing so might present only tentative ranges of options or inscrutable conclusions. And then, will we use other AI to evaluate the claims of the first AI output? Certainly, that option will be available, but should we trust AI to evaluate claims about what we should do? Normative advice draws on ethical values as well as factual claims about what is and predictions about what will be.

Thinking about progress will be increasingly affected by AI-generated claims and predictions, but we urge caution in relying on AI to be the silver bullet that solves all of our problems. Even if we manage to harness the power of generative AI to our advantage, we should still realize that our most pressing problems are multi-faceted and resistant to purely technological fixes.

As with our other technologies, AI should be treated as a tool that is in need of evaluative oversight. For this reason, it is imperative that we strive to understand how AI works, and especially what its limitations are. This is even more true if we are outsourcing to AI our judgments about what we should do.

But this doesn't mean we cannot also learn from generative AI, and this may be most evident in the way it writes. Current academic standards tend to require scholars to track down every relevant article that may exist on a topic, and reference them all. This has come at the expense of time and energy that could be devoted to fresh thinking. To be sure, some of the literature must be read, but it is increasingly difficult to read it all. In response, young scholars may be tempted to read only titles or abstracts, citing the papers without reading them. Generative AI tends to present ideas cleanly, without copious references interfering with readability. Experts may have to follow suit if they want their ideas to reach larger audiences without passing through the filter of AI assistants and essay writers that would likely smooth over the important particularities of the new idea.

On the other hand, one of the main points of difference between academic writing and AI-generated text is the accuracy of the citations. ChatGPT and other large language models, we have learned, have a proclivity to "hallucinate"—to confidently interweave facts with fiction. A generative AI that is asked how to treat a serious medical condition might produce some plausible-seeming text composed out of accurate medical advice, but with homeopathic and other quack remedies strewn throughout. When asked to support its claims, ChatGPT tends to fabricate citations to authentic-sounding academics, papers, and institutions that do not exist.

Nonetheless, within a year of ChatGPT's release in late 2022, some appreciable progress had been made on the problem of AI hallucination. If you were a traditional academic looking to justify your continued employment, you could still emphasize your advantage in accurate referencing, but you would be foolish to rely on that argument. Generative AI's referencing problem seems easily fixable.

After all, one reason that ChatGPT hallucinates academic references is that it lacks access to the large repositories of gold-standard scholarly research. As a general matter, the parts of the internet beyond the paywalls of academic journals do not offer particularly good coverage of many academic specialties. Thus, if pressed to cite accurate academic literature, ChatGPT will often draw on sources like Wikipedia, where many entries do indeed cite academic articles.

Given these artificial constraints, it is easy to imagine a future in which the problem of hallucinated citations has largely been addressed. Leading academic publishers like Taylor and Francis may themselves create a generative AI with special access to their archives. Not only would it be able to reference generations upon generations of work in the humanities. It also would be able to correct the occasional referencing error produced by a human author. What then for academic writing?

Academics should look to other areas in which humans have seemingly been overtaken by machines. In the case of AI, the most commonly cited examples are chess and the fiendishly complex strategy game Go. Garry Kasparov, the world chess champion from 1985 to 2000, has extensively documented his own experiences over a period when computers abruptly went from beating only mediocre players to easily beating the best ones. Yet it is worth noting that the loss of human dominance over the chessboard did not lead to a mass renunciation of the game. Rather, it required an adjustment in how humans play chess and what we value about it.

The ancient Chinese game Go was the next to go. For a while, human Go players accepted that machines might eventually achieve mastery of them, but they assumed that human capitulation was still a long way off. Reflecting pityingly on their chess counterparts, Go players took comfort in the greater complexity of their own game. According to one frequently cited figure, the number of legal board positions in Go is 2.1×10 to the power of 170 (Tromp and Farnebäck 2006), which is greater than the number of atoms in the observable universe.

Korean Go champion Lee Sedol's loss to DeepMind's AlphaGo in 2016 punctured this confidence about human supremacy. AlphaGo did not tackle Go by crunching through the 2.1×10 to the power of 170 possible positions, but it did work out how to approach the game's challenges in a way superior to any human. Looking ahead, humans will continue to enjoy Go, but they will have to negotiate a place for themselves in a world where they are assuredly not the best Go players. The continued flourishing of chess suggests that this is eminently doable, but the negotiation process will be interesting.

You can see a similar dynamic playing out among journalists, many of whom emphasize the deficiencies of generative AI. "Your correspondent's job is safe for now," they tell us, with a collective sigh of relief. Yet such assurances should be placed in the context of other stories about writing tasks once performed by journalists now being done by machines.

Obviously, it is a mistake to defend human scholarship by pointing to some advantage that may well vanish with future iterations of AI-written scholarship. The question is how to identify advantages that might be more enduring. What do human thinkers currently do that AIs most likely will not be able to do soon? When we engage in such self-inspection, we should be careful to avoid idealizing AI's capacities.

Consider Data, the endearing humanoid AI in *Star Trek: The Next Generation*. Data spends much of his time trying to be more human, and he can indeed do almost everything that humans can. He seems to be far and away the most valuable member of the starship's crew, and a minimal redesign would presumably equip him to replace human journalists or philosophers, too. Astute viewers thus find themselves often wondering why all the ship's crew hasn't been replaced by machines. Perhaps sensing this objection to the existence of the human crew, the show's screenwriters came up with a storyline in which we learn that Data and his evil counterpart, Lore, were limited productions. There simply is no factory to mass produce them.

In any case, we shouldn't be overly distracted by the AIs of our imagination. If Google abruptly begins churning out androids with Data's capacities, we are likely all screwed. But there is a big gap between Data and any AI we will get soon.

What matters for those of us living today is that we can engage with ideas about our own future better than any machine can. We don't have to out-think Data. We merely need to consider the ways in which we can be expected to outthink the AIs of the 2040s, and offer these as more durable bases for human self-esteem. Our ability to think deeply about how ethical values *should* interact with claims about how the world works and what it may become should set us apart for a while. Combining our normative proficiencies with human imagination itself may well be a secure resting place for our sense of purpose. We have tried to reflect that insight in our own thinking and writing about the future, both in the pages that follow and in our writing that is geared toward a general audience.

We hope that readers too will not give up thinking about what we can do to help progress our species safely into the future. Thinking about progress is too important to outsource to AI. Besides, if we leave it up to them, it might not be long before they lock Dave and the rest of us out of the spaceship.

Hamilton, New Zealand   Nicholas Agar
Cary, NC, USA   Stuart Whatley
Hamilton, New Zealand   Dan Weijers

References

British Academy. 2022. *The Teaching-Research Nexus: Project Summary*. London: The British Academy. https://doi.org/10.5871/hepolicy/teaching-researchnexus

Davidson, Cathy N. 2017. *The New Education: How to Revolutionize the University to Prepare Students for a World in Flux*. New York: Basic Books.

Tromp, John, and Farnebäck, Gunnar. 2006. Combinatorics of go. In *International Conference on Computers and Games*, pp. 84–99. Berlin, Heidelberg: Springer Berlin Heidelberg.

Contents

Chapter 1
Introduction

Contents

For most of human history, people, relying on agriculture, have prayed for rain. Today, we, relying on technology, pray for continued scientific breakthroughs. Whether we are seeking solutions to climate change and disease or pursuing lofty ambitions such as the colonization of Mars, even pessimists and skeptics tend to hold out hope that someone, somewhere, will come up with something sooner or later.

After all, that has been the modern creed at least since the Enlightenment. We are confident that there are no limits to human ingenuity. The doctrine of progress, the *fin de siècle* historian J.B. Bury observed, is one of the central ideas of Western civilization (Bury 1920, p. 1). And why shouldn't it be? Our commitment to the pursuit of innovation has been rewarded in the form of ever-higher material gains. It stands to reason that we would expect continued progress of this kind. Past performance does not guarantee future returns, of course, but nor can one judge too harshly our eagerness for ever more material progress. It is this anticipation that underpins the sprawling field of academic and pop futurism, where expert prognosticators of all stripes extrapolate from current trends to offer a portrait of what's next.

For centuries, we have convinced ourselves that every problem is solvable with enough brain power, particularly when applied through new technologies. And with the rapid development of COVID-19 vaccines, this conviction has received a new breath of life. Commentators have heralded a "Great Acceleration" and the onset of a "New Roaring '20s"—a reference to the decade of rapid economic growth that followed the influenza pandemic of 1918 (Pethokoukis 2021).

© The Author(s), under exclusive license to Springer Nature Switzerland AG 2024   1
N. Agar et al., *How to Think about Progress*, Library of Ethics and Applied
Philosophy 42, https://doi.org/10.1007/978-3-031-68938-3_1

Before the COVID-19 pandemic, the prevailing narrative had been one of disappointment, characterized by Peter Thiel's complaint that while we wanted flying cars, we got Twitter instead. Best-selling economists warned of a "Great Stagnation" and the fall of growth, owing to technology's inability to pack the same punch as it once did. But following the fast arrival of COVID-19 vaccines, many commentators began to welcome the dawn of a new era of progress in biotechnology, digital currencies, transportation, energy, and other fields (Macaes 2021).

Still, we as a global civilization are struggling to muster a sufficient response to challenges like climate change. With the arrival of COVID-19, we learned that big, societal problems don't patiently wait in line, giving us time to solve each in turn, while never confronting us with more than we can handle.

Pandemics and climate change can both be understood as *civilizational problems*, which is not the same thing as an existential threat. The latter implies human extinction. While COVID-19 was a nightmare, its total death toll barely put a dent in the global population. Similarly, climate change is very unlikely to eradicate humanity; even the worst-case scenarios do not envision Earth turning into Mercury, with its daytime temperatures of 430° Celsius. But pandemics and the climate crisis do both threaten our way of life on a global scale. And as we confront these problems, we should reflect on the attitudes that guide our responses to them.

Civilizational problems, by definition, affect or implicate everyone. They are the kinds of challenges that inspire a "wartime mobilization." By declaring war on things like poverty, climate change, cancer, and COVID-19, we set these issues apart from others. By setting out to colonize Mars, we tap into a much longer tradition of humankind's conquest over nature. But when thinking about civilizational problems—"Big Cs" like cancer, climate change, COVID-19, and colonizing Mars—we should be mindful that our secular faith in technological solutions is precisely that: a leap of faith. We are constantly waiting for the next big technological fix to arrive in the nick of time, because we implicitly accept that with enough information, practically anything is possible (Deutsch 2011, p. 56).

The Horizon Bias

Complicating the picture implied by such optimism is what we call the horizon bias: *the modern propensity to believe that anything we can envisage accomplishing with technology is therefore imminently in reach.* The horizon bias not only leads us to think that ambitious targets are closer than they really are; it also subjects us to an endless treadmill of dashed hopes. It is an enduring feature of an age in which futurism has become a booming industry.

As we will see, futurism is the offspring of the eighteenth-century doctrine of Progress. Early works in the genre, such as Sebastian Mercier's 1771 novel *L'An 2440*—one of the first works of utopian fiction set in the future—extrapolated from the technological advances of the present to consider what plausibly might come next. These Enlightenment-era works represented a new future-oriented consciousness—and a new, widespread sense of expectation for what industrialism would bring next.

Over the past 250 years, there have been ongoing efforts to turn the art of forecasting into something resembling a science. One of the most famous and influential examples of this was H. G. Wells's 1902 book *Anticipations of the Reaction of Mechanical and Scientific Progress upon Human Life and Thought*, which offered "a rough sketch of the coming time, a prospectus, as it were, of the joint undertakings of mankind in facing these impending years." Wells declared that he was going beyond the realm of futuristic fiction to offer "quite serious forecasts and inductions" (Wells 1902, p. 2).

Wells's project was continued in earnest after the massive upheavals of the two world wars. In another famous futurist work, the 1970 book *Future Shock*, Alvin and Heidi Toffler (though the latter was not originally listed as an author) set out to consider not only the technologies that were likely to come, but also their social and psychological implications (Toffler 1970, p. 5). The goal of the book was to make readers more future-conscious. This objective would appear to have been achieved. It is no exaggeration to say that the culture of Western advanced economies today harbors an obsession with the future.

Anticipations—and hype—about technology are a major engine of the news cycle and the foundation of a massive venture-capital industry. Futurism is the VC sector's bread and butter. VC firms like Prime Movers Lab now publish roadmaps to the year 2050, anticipating, among other things, commercial nuclear fusion plants operating by the 2030s; near-Earth asteroid mining by the 2040s; and the ability to slow or partially reverse human aging by 2050 (Prime Movers Lab 2021).

By focusing entirely on the pursuit of vaguely defined outcomes that technology might deliver, we end up on an endless journey—staring at an ever-receding point over the horizon, when we could be reflecting more deeply on our immediate surroundings. Moreover, by giving us too much confidence in our own technological abilities, the horizon bias leads us to exhibit a potentially dangerous preference for technological solutions to complex problems.

The horizon bias applies to our present and future, but it is rooted in a selective memory of our past. Technology's many headline successes—eradicating smallpox, putting a man on the moon—dwell permanently in our collective memory, offering strong inductive evidence for the power of human ingenuity. At the same time, we often forget (or are oblivious to) all the times that technology promised to solve some problem but didn't.

Remember the early years of the internet, when countless commentators and politicians from Al Gore to Newt Gingrich proclaimed a new age of freedom and democratization? The problem wasn't just that they were being naïve, notes historian Howard P. Segal, it was that they were effectively repeating the same claims made by many previous generations of optimists who looked forward to revolutions in communication and transportation technologies (Segal 2006, p. 66). Yet even if one accepts that "freedom" has expanded in some places over time, it is a stretch to argue that specific technologies—rather than, say, social and political developments—were the cause.

One reason for our serial forgetting is that we tend to fixate on certain key facts about new technologies, rather than the feelings that earlier instances of techno-hype might have incited in the public. In advanced, industrialized societies, school-children learn about the many inventions that made the modern age: the steam engine was followed by the internal combustion engine, which was followed by precision engineering, and so on. That story is simple and straightforward. More complicated is the story of our subjective experiences. We generally forget (or were not around to hear) what exactly was promised by those offering technological solutions at any given time in the past. With less memory or knowledge of all the hopes that were disappointed, we are left with a picture depicting only the hopes that were met.

To be sure, some past successes far exceeded the hopes of even their most ardent proponents—and these we remember fondly. Basic "low-hanging fruit" like flush toilets and indoor plumbing radically changed public health, allowing for an unprecedented increase in average lifespans. But, again, these subjective dynamics cut both ways. Plenty of other past successes might still qualify as such, but they would be viewed as disappointments by those who originally conceived them (as in Peter Thiel's famous observation that we wanted flying cars but got social media instead).

Indeed, this has been a recurring feature of the War on Cancer. Cancer researchers in the early 1970s were so confident in the trajectory of new chemotherapies that they expected the disease to be defeated within 5 years. They would be dismayed to see that these treatments have made hardly a dent on the overall cancer mortality statistics 50 years later. Insofar as they represent an improvement on what came before them, chemotherapies qualify as a success; yet against the specific criteria of what their earliest proponents promised, they are in that sense a failure.

Just as history is written by the victors, the story of technological progress features only the breakthroughs that actually panned out. But only by cutting out the warp and weft of subjective experiences can one arrive at the simple conclusion that Technological Man consistently accomplishes what he sets out to achieve.

By regularly forgetting the hopes that we (or previous generations) were originally encouraged to entertain with respect to technology, we are like those who frequent casinos. Casino operators have long known that to keep someone gambling, it is better to offer mystery prizes than prizes that are described in full (Binkley 2009). Since the mystery prize can be anything, it will on very rare occasions be something great, like a new Porsche. But when it turns out to be a free meal in the casino steak house, the gambler will soon forget his initial Porsche-level excitement. He'll take his steak and then rush back eagerly to listen for the next mystery prize.

The untold part our grand technological story is replete with visions of Porsches that never materialized, and then quickly faded from our memory. When it comes to our past hopes and expectations for technology, our memories are no more retentive than those of a dog pleading for a scrap of food, and with no apparent memory of the preceding five morsels she already received.

By remembering past *expectations* alongside past achievements, we can examine our own present hopes for technology more fully. In thinking about civilizational threats, technology obviously will have an important contribution to make. But insofar as we are led by the horizon bias to prioritize technology-driven solutions, we risk ignoring strategies in which new technologies play only a subsidiary role.

These biases matter. Over the course of his career, a racially biased police officer will doubtless arrest genuine criminals. But to the extent that he is biased, he is a worse cop, because he will be more likely to make wrongful arrests, and less likely to make arrests that he should make. The same applies to our bias in favor of technological solutions. Sometimes, we will be right to rely fully on technology. Yet over time, our preference will prevent us from even considering many possible responses to our most pressing problems.

By understanding this predilection, we can start to rebalance our responses to systemic threats, removing some of the marketing sheen from the technologies that we can see ourselves deploying against climate change and future pandemics. The temptation to imagine technological solutions to these problems can easily lead to overindulgence when we start fantasizing about what these technologies *could* be, and how they *could* work in the real world. This is Porsche-thinking in action.

For example, suppose we could enhance our immune systems to ward off any member of the coronavirus family, present or future. We could then declare victory even against the common cold. Or suppose we could create technologies that would efficiently and cheaply remove carbon-dioxide from the atmosphere with the turn of a switch. We could then maintain or even double down on our current way of life without having to worry about the climatic effects. Just thinking about such straightforward solutions offers a hit of dopamine. And when these technologies fail to materialize, we promptly forget about that source of hope as we line up for the next bold idea.

What's the problem with imagining a future in which technological advances fix our most pressing challenges? It is that the more time we spend obsessing over what technology could do for us, the less attention we give to social and political imaginaries. We tend to prefer engineered solutions because these make fewer awkward requests of us.

By contrast, social and political advances usually involve difficult changes that a significant cohort of people would avoid if given the option. Why think about what it would take to reduce meat consumption when we could be thinking about technologies to make ethical, low-carbon lab-grown burgers and steaks? Never mind that the availability of abundant ethical lab-grown meat would come with costs of its own; namely, by encouraging even more consumption of foods that are associated with a wide range of negative health outcomes.

The difficult scenarios that we would prefer not to think about each entail some level of self-sacrifice, trust in others, or old-fashioned hard work. It is perfectly understandable that we would seek to avoid these by ceding the floor to those offering free lunches. In the absence of technological fixes, the next-best thing we have is metaphor.

For example, we have an abiding tendency to wage metaphorical wars on hard problems. In 1971, US President Richard Nixon declared war on cancer, whereas today one often hears arguments for a wartime mobilization against climate change or COVID-19. The idea of declaring war on atmospheric carbon or on a microscopic viral particle seems quixotic. But there is a reason to retain this way of thinking, at least until our civilization acquires serviceable replacements. Wars have typically demanded organized responses, and marshaled the talents and ingenuity of the broader population.

Of course, Susan Sontag famously rejected the metaphorical war on cancer and hoped for new metaphors to direct our engagement with that disease (Sontag 1978, p. 677). But we see the war metaphor's frequent use as an indicator of its cultural power. If it can muster an organized resistance to climate disasters or pandemics without requiring too much explanation, we should avail ourselves of it. It is, in this sense, a powerful *ideological* technology—a widely understood idea that we have fashioned to make the most of our social natures and compacity to work together against common enemies.

In the final chapters of the book, we gesture toward less bellicose metaphors that may be available to carry out the same job. These metaphors could draw on our deep affiliative instincts rather than on our destructive urges. Perhaps a truly enlightened future age will look back on the use of war metaphors as primitive and passé. But, in the meantime, they are what we have to work with.

Moreover, war metaphors imply a collective response that differs from the pursuit of technological quick fixes. Technology has certainly played a key role in all past wars, but it has not been the whole story. When we recount the victory over fascism in World War II, we can focus on Spitfire fighter planes, Sherman tanks, and the A-bomb, but we tend to prefer the human stories, such as of the "Band of Brothers" who worked together and risked everything for their country and each other. These are good reminders of what it really takes to face down a truly hard problem. (If a problem can be solved solely with technology, it perhaps was not so hard to begin with.)

Infantry or Cavalry?

The lure of technological imaginaries featured prominently in the COVID-19 crisis. Globally, there was a clear divide between countries that mustered effective *social* responses to the pandemic, and countries that did not. Political leaders in the first group offered straightforward directives to mandate lockdowns and face masks— two scientifically proven measures to reduce transmission—whereas those in the second group hesitated or outright refused to impose lockdowns, out of concern for the short-term economic costs. Predictably, many countries in the second group suffered much higher rates of infection and death.

There is no simple, single-factor explanation for the differences between countries. However, we would submit that at least part of it can be explained by a bias toward technological solutions among those in the second group. The more successful countries did a better job of resisting free-lunch thinking, and thus readily accepted the need for masks, social distancing, lockdowns, and other inconveniences. The others focused more on the technological breakthroughs with the potential to defeat the virus singlehandedly. If there is one achievement that even Donald Trump's harshest critics are willing to acknowledge, it is that his administration created Operation Warp Speed, which played a key role in delivering a vaccine.

To be sure, the countries that managed the virus better than others used technology, too, by coding mass testing and loading smartphones with tracking software to carry out contact-tracing operations. But the technologies in these cases tend to be supplemental, rather than primary, elements of the broader strategy.

Early on in the pandemic, New Zealand Prime Minister Jacinda Ardern proposed that Kiwis track their locations and contacts using the "homespun" technology of pen and paper. The fact that we have smartphones capable of doing this more efficiently is largely incidental to what really mattered for the national response to work: social solidarity and a widespread willingness on the part of citizens to keep track of where they had been and whom they had come into contact with (SBS News 2020).

Now consider the US. In November 2020, Anthony Fauci, then one of the lead members of Trump's White House Coronavirus Task Force, responded to good news about the development of vaccines by remarking that "the cavalry is coming." He hastened to add that Americans should still continue to wear masks, avoid crowds, stay 6 feet apart, and wash their hands frequently (Guzman 2020).

But Americans' behavior did not improve. On the contrary, millions traveled and got together with friends and family for the holidays. In mid-November 2020, the US was averaging around 150,000 new infections per day; by January 8, that figured had grown to almost 260,000—its highest peak before declining sharply with the rollout of vaccines (New York Times 2021).

In other words, many Americans' tacit response to Fauci in November was, "Thanks for the warning Anthony, but you had us at 'vaccine!'" Here was the perfect technological solution to the problem. The new mRNA vaccines proved even more effective than many had imagined. For once, the mystery prize really was a Porsche. If Americans could just get those shots into their arms fast enough, there would be no need to continue with the unpleasantries of the pandemic. The hope now is that SARS-CoV-2 will become a nasty historical curiosity alongside the influenza pandemics of 1918 or 2009. If we can all get vaccinated, we can forget about the annus horribilis of 2020.

It currently does not seem likely, but if the vaccines do make SARS-CoV-2 extinct, there is a good chance that future generations will remember COVID-19 as a bad pandemic that was fixed with the technological marvel of mRNA vaccines. The millions who died from the virus certainly won't be around to point out that the story was more complicated than that. If future generations are as biased toward technological solutions as we are, they will happily overlook the fact that if a civilization's survival depended on the calvary riding in, it must not have been doing a very good job of defending itself.

The Lure of the Climate Hack

One can also see the effects of our tech-oriented preferences in proposed responses to the climate crisis. For example, in *How to Avoid a Climate Disaster: The Solutions We Have and the Breakthroughs We Need*, Bill Gates, the Microsoft co-founder and philanthropist, explains that we will need technological breakthroughs to reduce carbon emissions to net-zero in time to prevent catastrophic climate scenarios (Gates 2021). But this outlook raises the crucial question of what we ought to do in the meantime. Would we be justified in simply biding our time, waiting for the calvary to arrive? Or do we owe it to our children and grandchildren's generations to act as if those breakthroughs will come too late, or never at all?

Eager for technological solutions, readers' minds will readily alight on some of the exciting technological possibilities that Gates describes. It feels good to join Gates in "geeking out" on the science of the issue. Whether he is exploring the feasibility of direct air capture (DAC) or solar geoengineering, we are drawn to any shiny technological *possibilia* that would spare us from painful or inconvenient acts of self-abnegation. Just think, with enough brain power and ingenuity, we could create instruments to nullify the negative effects of fossil fuels, allowing us to continue burning these cheap energy sources indefinitely.

Consider DAC technologies that remove carbon from the air. Some power plants already have "point capture" mechanisms that catch carbon as it is produced; but these are expensive and inefficient. As a technological imaginary, DAC is the far superior option, since it doesn't need to be affixed to a power plant or present at the time the atmospheric carbon is emitted. Indeed, it theoretically has the potential to capture the carbon that was first belched from those coal-fired plants that drove the first Industrial Revolution.

For his part, Gates is careful to distinguish between the technologies that we actually possess and the technologies that we may develop at some point in the future. He offers DAC as merely a thought experiment, and cautions that even if it can be scaled up globally, that process would almost certainly not happen fast enough to spare the climate from serious harm.

Plenty of other contemporary tech futurists would say that Gates himself is too pessimistic. They point to the exponential advances in the digital realm to suggest that seemingly utopian technologies are much closer than we think. Thus, while Gates at least accepts that we cannot bet on future technologies coming to our rescue, those who are more tempted by technological solutions will question whether we really need to go through all of the irksome self-denial of decarbonizing the economy. Even those not normally embracing technological solutions might find faith in so-called green tech the best way to deal with their climate anxiety (Weijers and Agar 2023).

It is no coincidence that major oil and gas giants like Shell now frequently tout DAC as the climate solution we've been waiting for. But the fossil-fuel industry is hardly alone. As James Temple notes, companies and governments have been racing to unveil seemingly ambitious net-zero emissions roadmaps that rely heavily on carbon removal, and an emerging start-up industry is leveraging this demand to attract investors, often with outlandish claims about what their technologies will do (Temple 2021).

Indeed, the outright denial of climate science has ironically given way to an unconditional faith in science's capacity to deliver technological breakthroughs just in the nick of time. Surely, the argument goes, it would be better to redirect some chunk of the profits from fossil-fuel consumption toward an innovation "moonshot" to achieve the necessary breakthrough in DAC. That would allow us to go green and clean simply by tidying up after ourselves—and all without ever renouncing carbon.

This attitude was on full display in Australia's response to the unprecedentedly damaging 2019–2020 bushfire season, which burned more than 18 million hectares of land (Phillips 2020). The usual opponents of climate action had difficulty maintaining their traditional strategies of denial, since the evidence for serious anthropogenic climate change had suddenly become so visible and obvious. But faced with such a harrowing scene, the Australian coal industry's defenders didn't renounce their advocacy of fossil fuels. They simply switched strategies, arguing that it would be pointless for a relatively small global emitter like Australia to start slashing its fossil-fuel use now. At this late stage in the game, the only sensible option is to come up with a technological fix that will allow everyone to continue on much as before.

The Big Cs

In the following chapters, we will explore our preference for technological solutions through the lens of a civilizational problem that has *not* yielded to human ingenuity or technological breakthroughs: cancer. Our broader discussion of technological progress—and what we expect from it—will revolve around this problem and the other Big Cs, like the favored quest of today's tech tycoons: colonizing Mars.

In Chapters three and four, we will focus on cancer, because it stands out as a powerful counterexample to weigh against the modern narrative of inevitable just-in-time progress, and on the quest to colonize Mars, because it offers a real-time look at the horizon bias in action. The history of our struggle against the disease illustrates how the horizon bias and our preference for technological solutions can cloud our thinking about truly big problems, understood as the kind of issues that we "declare war" on.

When US President Richard Nixon declared war on cancer in 1971, he and his advisers expected that victory would be imminent, with a "cure" arriving within 5 years (Surh 2021, p. 219). Since then, cancer researchers, politicians, and technologists have repeatedly promised "the cure," even though that outcome remains fully out of our reach.

For a culture and society that prides itself on its mastery over nature, cancer seems eminently curable, because it seems like a problem that is easy to characterize. It is essentially a disease of unregulated cell growth, and we have known how to kill cells ever since the discovery of fire. If the challenge is to kill the unwanted cancer cells while leaving healthy cells intact, then, surely, we just need to come up with the right tools.

When we characterize cancer in this way, it seems curable, because we have opened the door for technological imaginaries. Faced with an engineering problem, someone just needs to discover a compound or invent a method that distinguishes the bad cells from the good ones. But, of course, if it was that straightforward, we would have already reduced the cancer death rate much more than we have.

This is not to suggest that we have made no progress against cancer. On the contrary, there has never been a better time in history to receive a cancer diagnosis. Given the choice, everyone would pick today's medicine over the medicine of the 1970s. But it is worth remembering what we were promised back then: *the* cure for cancer—not just some quasi-cures for some types of cancers (something doctors have occasionally been able to achieve for centuries).

Victory in the War on Cancer was supposedly imminent, and yet nothing that we would recognize as "the cure" has ever materialized. We get frequent surges of dopamine each time some new breakthrough is announced, but that is only because we tend to forget all of the false dawns that we have already lived through. In 1998, it seemed apparent to DNA co-discoverer James Watson that new drugs designed to deprive tumors of blood would cure cancer in 2 years. Now, similar hopes are being placed in new mRNA-based cancer treatments.

Is this time different, or are we once again grasping at a point on the horizon that remains out of reach, even though we can see it?

References

Binkley, Christina. 2009. *Winner Takes All: Steve Wynn, Kirk Kerkorian, Gary Loveman, and the Race to Own Las Vegas*. Hachette Books.
Bury, J.B. 1920/2006. *The Idea of Progress: An Inquiry Into its Origin and Growth*. BiblioBazaar.
Deutsch, David. 2011. *The Beginning of Infinity: Explanations That Transform the World*. Penguin Books.
Gates, Bill. 2021. *How to Avoid a Climate Disaster: The Solutions We Have and the Breakthroughs We Need*. Knopf Publishing Group.
Guzman, Joseph. 2020. Fauci: 'The Cavalry Is Coming' with Coronavirus Vaccine, But Public Health Measures Still Needed. *The Hill*, 12 November.
Macaes, Bruno. 2021. After Covid, Get Ready for the Great Acceleration. *The Spectator*, 13 March.
New York Times. 2021. Coronavirus in the US: Latest Map and Case Count. *New York Times*. https://www.nytimes.com/interactive/2021/us/covid-cases.html. Accessed 25 Apr 2021.
Pethokoukis, James. 2021. Biden and the New Roaring '20s; supersonic airliners; the 21st century slowdown Musk, Grimes and AI; bullet trains; and more… *Faster Please*. Substack newsletter, 8 June.
Phillips, Nicky. 2020. Climate Change Made Australia's Devastating Fire Season More Likely. *Nature* 4 March. https://doi.org/10.1038/d41586-020-00627-y.
Prime Movers Lab. 2021. *Breakthrough Science Roadmap: Life in 2050*. https://www.primemoverslab.com/roadmap/. Accessed 12 Sept 2021.

SBS News. 2020. Jacinda Ardern Wants New Zealanders to Keep a Diary to Help with Coronavirus Contact Tracing. *SBS News*, 19 April.

Segal, Howard P. 2006. *Technology and Utopia*. American Historical Association.

Sontag, Susan. 1978. Illness as Metaphor. In *Sontag Essays of the 1960s & 70s*. Library of America.

Surh, Young-Joon. 2021. The 50-Year War on Cancer Revisited: Should We Continue to Fight the Enemy Within? *Journal of Cancer Prevention*. Dec 30 26 (4): 219–223. https://doi.org/10.15430/JCP.2021.26.4.219.

Temple, James. 2021 Carbon Removal Hype is Becoming a Dangerous Distraction. *MIT Technology Review*, 8 July 8.

Toffler, Alvin. 1970. *Future Shock*. New York: Random House.

Weijers, Dan, and Nicholas Agar. 2023. Why We're Seduced by Climate Tech and What it Means for Our Happiness. *Frontiers in Climate* 5: 1193581. https://doi.org/10.3389/fclim.2023.1193581.

Wells, H.G. 1902. *Anticipations of the Reaction of Mechanical and Scientific Progress Upon Human Life and Thought*. 2nd ed. London: Chapman & Hall, LD.

Chapter 2
The Rise of the Futurists

Contents

In early 2022, Intelligence Squared US hosted a debate on whether gene editing should be used to improve our offspring. Among the chief concerns associated with this powerful new biotechnology is that it will lead to a future in which the "haves" will be able to improve their children's genetic profiles in ways that are unaffordable to the "have nots." But Amy Webb, the CEO of the Future Today Institute, argued that "there is no evidence to support the claim that gene editing will benefit the wealthy specifically" or "necessarily take a dystopian route." When confronted with the fact that access to basic health care and reproductive technologies like in vitro fertilization is already glaringly inequitable in the United States, Webb urged the audience "not to use the past to predetermine what the future might look like" (Webb 2022).

A journalist-turned-corporate consultant and self-described "quantitative futurist," Webb relies on data and next-order-outcome models to justify her confident assurances of how a technique like gene editing is or isn't likely to be used in the future. Putting aside the fact that Webb's prognostications also must use "past" data (for there is no other kind), what is the non-expert listener to make of such claims? When purported experts with an abundance of seemingly relevant information tell us what kind of future technological or social developments we should prepare for, should we give them the benefit of the doubt?

On the one hand, forecasting is a standard practice in many fields. Demographers regularly offer projections of future population figures without stirring too much controversy, and climate scientists deploy sophisticated models to determine the probabilities of various long-term climate scenarios. Major financial institutions like the International Monetary Fund and the World Bank generally do a respectable job projecting broad metrics like annual global GDP figures (making regular revisions to account for new conditions, like a pandemic or a major war).

On the other hand, Webb's prognostications are the products of a futurism industry that caters to businesses' abiding fear of uncertainty by offering apparently scientific "strategic foresight" on any subject for which there is a paying subscriber (FTI 2022a, p. 23). As such, she has an interest in presenting anticipations about the future in ways that seem closer to knowledge than to mere opinion, even when addressing empirical questions that are subject to an untold number of variables and contingencies (like the one above about gene editing).

In the news-media industry where Webb cut her teeth, a familiar adage is that while fact-finding and investigative reportage is expensive to produce, opinions come cheap. For futurists to have a viable business model, they must offer something that seems like more than mere commentary. They must offer scientific rigor—or at least a pretense of it—when extrapolating from present "trends" to describe various "plausible futures."

Hence, the Future Today Institute's flagship annual *Tech Trends Report* describes a methodology consisting of a multi-step forecasting model and scenario-mapping process that considers four different time horizons (12–24 months, 2–5 years, 5–10 years, 10+ years) and almost a dozen sources of large-scale change, all to identify genuine trends (as opposed to passing fads) (FTI 2022a, p. 9–13, 17).

What insights does this approach yield? For the most part, readers are furnished with a mix of observations about the present and recent past (such as, "the pandemic exposes the fragility of our global supply chain"; artificial intelligence will automate some tasks and augment others) and hype that may or may not augur materially significant changes within the next 10 years (FTI 2022c, p. 21; 2022b, p. 7).

The Future Today Institute's 2022d report, for example, states that "synthetic biology will make aging a treatable pathology" (FTI 2022d, p. 7). As evidence for this claim, it points to "Columbia University researchers [who] discovered that it might be possible to record and store information about cells as they age," and to work by Harvard synthetic biologist George Church (Webb's Intelligence Squared US debate partner) in which gene therapies seemed to be effective at reversing "obesity and diabetes while also improving kidney and heart function" in mice (FTI 2022d, p. 20).

However, since the report prudently avoids offering any timeline for *when* aging will become a "treatable pathology," it will be difficult to test the validity—or at least the precision—of the claim. Nor is it easy to assess the Future Today Institute's previous prognostications. Even upon request, it does not make its past *Tech Trends Reports* available.

Whether one considers futurism to be an art, a science, or merely a glorified form of market research, there is little question that the discipline has undergone a full and rapid process of professionalization over the past half-century or so. Various "futures studies" theories and methodologies have been formalized, frameworks for assessing "foresight competency" have been introduced, and futurists have increasingly adopted a shared body of jargon. They have also been at pains to avoid predictions, focusing instead on trend-spotting and scenario-mapping. The point of futurism is not to predict, futurists insist, but to describe alternative outcomes and then offer forecasts.

However, since statements about the future (or contingent possible, probable, and preferable futures) come with their territory, futurists' efforts to eschew strong predictive claims are not always successful, especially when such claims attract paying customers. Indeed, as we will see in later chapters, some of the world's most well-known futurists, like Ray Kurzweil, do not shy away from offering ambitious predictions.

New Vistas

The twentieth-century bibliographer I.F. Clarke traces the roots of modern futurism as far back as the thirteenth century, when the medieval monk Roger Bacon foresaw that the deepening of scientific knowledge could lead to self-propelled planes, trains, and automobiles. In or around 1260, Bacon mused that: "Machines for navigation can be made without rowers so that the largest ships on rivers or seas will be moved by a single man in charge with greater velocity than if they were full of men. Also cars can be made so that without animals they will move with unbelievable rapidity ... also flying machines can be constructed so that a man sits in the midst of the machine revolving some engine by which artificial wings are made to beat the air like a flying bird" (Clarke 1979, p. 3).

Such thinking was extraordinary and novel for the time, and it would remain in the cloisters for another three centuries. For most of history, anticipations about the future had assumed that human affairs followed an eternal cyclical pattern of growth and decay—as indeed they did for agrarian societies and the many empires that had risen and then fallen. Even in groundbreaking speculative works like Thomas More's *Utopia* (1516), the imagined society existed not in a different time but merely in a different place; moreover, it owed its achievements more to social and cultural arrangements than to novel technologies (Segal 2006. p. 18).

In Clarke's view, the widespread sense that the future would be fundamentally different from either the past or the present did not emerge until the age of industrialization, with books like Sebastien Mercier's 1771 utopian novel, *L'An 2440*, which signaled the advent of not just a new literary genre but a new worldview.

With numerous editions and translations in the two decades after it first appeared, Mercier's work of speculative prognostication was a wild commercial success. Channeling his era's faith in technology-driven progress, he described a future of peace and social harmony, governed by philosopher-kings. Falling asleep in the 1770s, his narrator wakes up in the twenty-fifth century to find a world where slavery has been abolished, the criminal-justice system reformed, and medicine subjected to science-based rationality. The people's clothes are made for comfort rather than pomp and circumstance, and education has become a public service provided by the state (with students no longer being expected to learn Latin or Greek).

On many counts, Mercier proved to be a competent futurist. Yet it is an open question whether his accurate forecasts were a matter of luck, common sense, or "strategic foresight." Most likely, they were a mix of all the above. The more trends that one identifies, the more likely that at least some will be borne out (in 2022, the Future Today Institute identified more than 500 trends in technology and science).

With that contextual point in mind, it is also worth noting that Mercier antici- pated many other developments that will sound absurd to modern readers. For example, he envisaged the territory of North America being returned to its original inhabitants, and he thought that Portugal might become a part of the United Kingdom. In his future, taxes, standing armies, and even coffee have all been abolished.

Had Mercier been a corporate consultant, it is unclear whether his clients would have been better prepared for various future scenarios than their competitors. Had they acted on all of his prognostications, they might have bet against the coffee trade just as it was about to gain a permanent market foothold in North America. Following the Boston Tea Party in 1773, coffee decisively replaced tea as the preferred drink in the New World.

Like today, Mercier lived at a time when each new scientific and technological breakthrough would incite breathless predictions about even loftier achievements to come. In the 1780s, early successes with balloon flight led the French encyclopedist Denis Diderot to predict—accurately—that humankind would eventually make it to the moon. A contemporary poet went further: "Do not talk about impossibilities, for nothing is impossible to determined effort." If the Montgolfier brothers' balloons can fly to the heavens, "open Hell to me, I'll put out the fires" (Clarke 1979, p. 30–31).

From then on, each generation brought a new host of what Clarke calls "profes- sional horizon-watchers." Technological innovation had made predictions common, and though earlier practitioners' techniques were nowhere close to as sophisticated as those used by futurists today, their basic method was the same: by extrapolating from the latest breakthroughs, they envisioned new realms of plausibility.

One of the most well-known practitioners of this method was the French philoso- pher Nicolas de Condorcet, who offered a vision of the society of the future in his posthumously published *Outlines of an Historical View of the Progress of the Human Mind*. Writing while in hiding from an arrest warrant at the height of the French Revolution, Condorcet averred that humankind can "foresee, with consider- able probability, future appearances" by extrapolating from the historical trends that created the present moment. Owing to the progress of the natural sciences, he con- jectured that the future would bring "the destruction of inequality between different nations; the progress of equality in one and the same nation; and lastly, the real improvement of man" (Condorcet 1795, p. 352–253).

Condorcet anticipated that the liberal principles enshrined in the new French constitution would continue to spread, defying any attempts by tyrants and priests to keep the public ignorant of them. All European colonies would soon enough achieve independence, he wrote, and the cultivation of sugarcane in Africa would "put an end to the shameful robbery by which, for two centuries, that country has been depopulated and depraved." State-backed monopolies like the Dutch East India Company would soon fall, and governments would introduce public education for the young and social-security schemes to support the elderly with funds paid into a "common stock" (Condorcet 1795, p. 358, 369, 371).

Condorcet also anticipated that economic productivity would steadily increase, allowing for a "smaller portion of ground [soil]" to "produce a proportion of provisions of higher value or greater utility." As people grew more enlightened, he surmised that war would eventually come to be regarded as the "most dreadful of all calamities, the most terrible of all crimes." He foresaw vast improvements in humanity's ability to compile and analyze information. He expected that scientific progress would deliver advances in the "art of instruction," which would deliver still more scientific advances. Such progress, he reasoned, could not fail to improve the art of medicine, prolong the average lifespan, and improve the human stock through selective breeding (Condorcet 1795, p. 383, 397, 401, 403, 410, 412).

Eventually, these forecasts were borne out to varying degrees. But, as one would expect, Condorcet also got much wrong about the future. He predicted the "destruction of inequality between different nations" and the "progress of equality in one and the same nation." He expected that with the fall of their monopoly chartered companies, Europeans would soon begin to respect other societies' independence. He looked forward to the day when the world would comprise only "free nations," when tyrants and priests would be consigned to history. He surmised that with the spread of reason, people would no longer be dazzled by genius, and "all the causes which produce, envenom, and perpetuate national animosities, will one by one disappear, and will no more furnish to warlike insanity either fuel or pretext" (Condorcet 1795, p. 353, 359, 364, 374, 398).

Condorcet also offered a prediction that one still hears to this day: that the march of medical science will put an end to "contagious disorders, as well as to those general maladies resulting from climate, aliments, and the nature of certain occupations," as well as "almost every other malady of which it is probable we shall hereafter discover the most remote causes" (Condorcet 1795, p. 410).

To be sure, not everyone was as optimistic as Mercier or Condorcet. Perhaps most famously, Mary Shelley's *Frankenstein* showed that advances in science brought not just new technological possibilities but also new ethical quandaries. Her account of a brilliant scientist who engineers a new life form in secret (and who also wanted to make aging a treatable pathology) has eerie parallels to today (Shelley 1992, p. 31). In 2018, the Chinese biophysicist, He Jiankui, announced that he had used the CRISPR gene-editing technology to create the world's first genetically edited babies—a revelation that led to his being fired, prosecuted, and sentenced to 3 years in prison (Greely 2019).

But Shelley was an exception. Most of the nineteenth century's proto-futurists were breathlessly enthusiastic about where they saw humanity heading. They were prone to the prevailing biases of their time, not least a deep faith in the inherently beneficial nature of "Progress." In their view, new technologies and ideas could not fail to move humankind ever closer to perfectibility. Steamships and railroads promised "to draw all nations into more intimate connexion and to convert the whole human race into one society," wrote Michael Angelo Garvey in his 1852 book, *The Silent Revolution, or the Future Effects of Steam and Electricity Upon Mankind*. A decade before the American Civil War, he anticipated that:

> The hour is coming, hastening with the momentum of ages, when the … argument of murder, rapine, famine and pestilence – shall be banished from amongst men, consigned to the chamber of horrors, in which history preserves the memorials of crime. … A new light is dawning upon the world which will render such monstrous deceptions impossible men, by a more perfect association with one another, and by the more perfect blending of their real interest which must ensue, will see in all the broadness and perspicuity of truth that the clumsiest and least rational way of disposing of enemies is to kill them. (Clarke 1979, p. 80-81)

Lending further momentum to such beliefs, Charles Darwin's *Origins of Species* appeared in 1859, revolutionizing humanity's understanding of its own past and the natural world more broadly. Among the future implications of his "theory of descent with modification through natural selection," Darwin suggested, were that humanity could "look with some confidence to a secure future … And as natural selection works solely by and for the good of each being, all corporeal and mental endowments will tend to progress towards perfection" (Darwin 1859, Chapter XV).

Great Anticipations

By the 1860s, futurism had become a lucrative enterprise for those who could meet the public's demand for savvy and imaginative insights into the technological frontier. None were more successful than the French author Jules Verne, widely considered one of the founding fathers of science fiction (alongside Shelley, its founding mother). Owing to the arrival of the telegraph, Verne and his publisher, P.J. Hetzel, were able to serialize and syndicate the 1865 book *From the Earth to the Moon* to newspapers around the world (Clarke 1979, p. 93). Within just months of the book's publication, Clarke tells us, "American publishers were negotiating with Hetzel for the translation rights, and within 5 years, Jules Verne had become one of the best known of French writers," not just among young people but across the wider reading population.

Key to Verne's success was that he went to great lengths to explain the precise workings of the technologies that featured in his stories, lending his futuristic accounts a patina of plausibility. For the next half-century, the reading public's demand for such scenario-mapping grew and grew, reflecting a general belief that science and technology eventually would produce a solution to every problem. As we will see in a later chapter, the ability to offer "how-possibly" explanations for future scenarios is a powerful technique for capturing the imagination of the lay public—and sometimes for fooling oneself into embracing an unjustified degree of optimism about future technological progress.

Following Verne, the next globally successful exponent of the futurist persuasion was the English author H.G. Wells, starting with the publication of his novella *The*

Time Machine in 1895. Many of today's futurists (including Webb) trace their own professional lineage to Wells, owing to his efforts to elevate "predictive writing" to the status of an academic specialty. As the twentieth-century futurist and Wells biographer W. Warren Wagar notes, Wells embodied the trends in proto-futurism that had preceded him and pioneered a methodological approach that is still useful and relevant to futurists today (Wager 1983).

Like Verne, Wells had both a fertile imagination and a comprehensive familiarity with the latest technological and scientific breakthroughs of his time. He also recognized the need for a more methodological approach to the "discovery of the future" (Wells 1913). Born in 1866 to a lower-middle-class English household, Wells's active mind allowed him to escape the life of a draper's apprentice. In 1884, he won a scholarship to the South Kensington Normal School of science, where he studied under the biologist T.H. Huxley, one of the era's great popularizers of Darwinism (Luckhurst 2017, p. xix).

When Wells began writing his works of "scientific romance" in the 1890s, he met with a wide audience because he had managed to combine rich, dramatic storytelling about the future with a genuine command of the cutting edge of scientific knowledge (particularly biology). Of course, this approach had its critics, with G.K. Chesterton mocking the "prophets of the twentieth century," each of whom offered a different account of "what would happen in the next age." To Chesterton, the new method of divination was quite simple: "They took something or other that was certainly going on in their time, and then said that it would go on more and more until something extraordinary happened" (Chesterton 1904, p. 4).

Chesterton was referring to Wells's non-fiction work of prognostication, *Anticipations of the Reactions of Mechanical and Scientific Progress upon Human Life and Thought*, which had first appeared, to wide acclaim, as a series of articles in the *Fortnightly Review*. Like Mercier and Condorcet before him, Wells got some things right and some things quite wrong.

Anticipating the inventions of the tank and the airplane, he offered a canny analysis of how the weapons of the industrial age would change the nature of warfare, leading to catastrophic world wars. He predicted the phenomenon of modern highway systems and suburban sprawl, as well as the emergence of labor-saving household appliances like dishwashers. He foresaw that English would become a global language. Later, in 1933, he famously envisaged war breaking out between Germany and Poland in 1940 (Hitler invaded Poland in September of 1939). And he imagined the creation of a universal, multilingual encyclopedia that would be constantly updated and revised in real-time. Unlike Wikipedia, however, Wells expected that this task would require a central committee (Wager 1983).

Mixed in with these prescient insights were prognostications that missed the mark. Wells expected the populations of London, St. Petersburg, and Berlin to exceed 20 million, and those of New York, Chicago, Philadelphia, and Hankow (what is now called Wuhan) to reach 40 million (all huge overestimates) (Wells 1902, p. 48). And more notably, he expected that both capitalism and the nation state would collapse before the end of the twentieth century (Wager 1983). Still, his ultimate contribution was to stoke his readers' imaginations and to vastly expand the realm of what people considered possible. In doing so, he had a vast influence on the future of futurism itself.

Wells's philosophy is encapsulated in his 1902 lecture "The Discovery of the Future," where he contends that "in absolute fact the future is just as fixed and determinate, just as settled and inevitable, just as possible a matter of knowledge as the past." By this he meant not that the future is wholly knowable, but rather that it is just as open to inquiry as is prehistory—a domain where "we are inclined to overestimate our certainty." Advances in archeology, geology, and other fields, Wells observed, had allowed scientists to assert facts about "the form and the habits of creatures that no living being has ever met, that no human eye has ever regarded, and the character of scenery that no man has ever seen or can ever possibly see." They owed these discoveries not to revelation but to a "new and keener habit of inquiry," a "new and more critical way of looking at things" (Wells 1913, p. 23, 28, 30).

With advances in science and methodology having opened a window onto the "inductive past," Wells believed "quite firmly that an inductive knowledge of a great number of things in the future is becoming a human possibility." The test of a sound scientific theory, he observed, lay in its ability to offer confident and reliable forecasts, such that even "so unscientific a science as economics" had gotten into the business. "If the specialist in each science is in fact doing his best now to prophecy within the limits of his field, what is there to stand in the way of our building up this growing body of forecast into an ordered picture of the future that will be just as certain, just as strictly science, and perhaps just as detailed as the picture that has been built up within the last hundred years of the geological past?" (Wells 1913, p. 37).

The Need to Know

With the arrival of the kind of total wars that Wells had anticipated, projects to foresee and plan for the future were taken up in earnest (Bell 2017, Chap. 1). In the 1920s, the Soviet Union adopted five-year plans intended to imagine a radically different future and then make it happen. In the United States, President Herbert Hoover established a President's Research Committee on Social Trends under the leadership of sociologist William F. Ogburn. Ogburn then went on to work for the Roosevelt administration, where he helped produce the 1937 report *Technological Trends and National Policy, Including the Social Implications of New Inventions*.

The upheavals of the first half of the twentieth century created an urgent demand for technocratic planning, giving rise to "operations research" and, with it, the modern think tank. Operations research sought to describe the behavior of complex systems, devise theories for understanding said behavior, and then offer foresight into how a system would respond to various changes (Dickson 1971, p. 22). After World War II, the civilian proto-futurist research efforts that had taken root during the war were continued through the creation of Project RAND (short for "research and development"), which embarked on a study exploring the *Preliminary Design of an Experimental World-Circling spaceship*—presciently anticipating the development of satellite technology before it had emerged.

Think tanks continued to proliferate in the early post-war decades, with the Palo Alto-based Institute for the Future emerging as the first self-identified futurist institution of its kind, in 1968. Owing to demand from government and corporate sponsors, many of these organizations focused heavily on forecasting and devising more robust methodologies for anticipating—and determining—the shape of things to come. Standard methods in the new research included trend extrapolation, relevance trees, and new forms of statistical modeling. Among the more notable and influential approaches developed were RAND's Delphi technique and Hudson Institute Founder Herman Kahn's school of scenario mapping (Dickson 1971, p. 35, 327, 333).

In the case of the Delphi technique, a wide range of experts are queried on a specific question and then furnished with each other's anonymized responses, allowing them to revise their views until something like an intersubjective consensus emerges (Dickson 1971, p. 329). At the Hudson Institute, Kahn helped to bring futurist thinking to the general public, prompting more people to start imagining what had hitherto seemed like unimaginable scenarios. In his groundbreaking 1960 book, *On Thermonuclear War*, he offered an exhaustive overview of various nuclear-deterrence strategies and the risks they entailed.

Kahn wrote his book with the goal of foreseeing and thus preventing future crises. He wanted his readers to open their minds to disastrous possibilities that they had not even considered, by imagining all the ways that things could go wrong to lead either to an intentional or an accidental initiation of nuclear war. In retrospect, Kahn's warnings could not have been more timely: 2 years later came the Cuban Missile Crisis.

The same year that Kahn published *On Thermonuclear War*, the French futurist Bertrand de Jouvenel launched the Association Internationale de Futuribles in Paris, with support from the Ford Foundation (Bell 2017, Chap. 1). The organization's initial goal was to strengthen and improve the public standing of forecasting, on the belief that the social sciences more generally should adopt a future-oriented mindset. Then, in 1964, Jouvenel published *The Art of Conjecture*, which sought to systematically improve the natural process by which we all transform knowledge of the past (*facta*) into images of the future (*futura*) (de Jouvenel 1967, p. viii).

Jouvenel made clear that when futurists speak of forecasts, they are referring to informed opinions, not facts or hard predictions that would lead listeners to assume that there is a science of the future. Futurists, he advised, should concern themselves with "futuribles": a "future state of affairs" whose realization "from the present state of affairs is plausible and imaginable." Though even ancient societies thought that human flight should be possible, aviation did not become a futurible until "certain new facts made its development conceivable." Thus, for Jouvenel, imminent plausibility served as a guardrail for the imagination; but, as we will see in later chapters, the same attribute can also become a slippery slope for believing overly confident predictions about future technological breakthroughs (de Jouvenel 1967, p. 16–18).

Finally, while most current futurists trace their lineage through Kahn and Jouvenel, we would be remiss not to mention Alvin Toffler, who did even more than the other post-war futurists to bring the evolving discipline to a mass audience. In his 1970 bestseller, *Future Shock*, Toffler offered a "broad new theory of adaptation" for an age of accelerating technological, social, political, and psychological change. Echoing Wells, Toffler also aimed to demonstrate a method for discerning the future: "Previously, men studied the past to shed light on the present. I have turned the time-mirror around, convinced that a coherent image of the future can also shower us with valuable insights into today" (Toffler 1970, p. 5).

Inspired by the more well-known concept of culture shock (the experience one feels upon suddenly arriving in an alien social environment), Toffler had coined the term "future shock" in a 1965 article to describe the psychological distress that comes with rapid, monumental change. One of the best ways to cope, he believed, was to adopt more of a future-oriented perspective, so that we are not constantly caught off guard by each new society-altering trend or development.

In the half-century since *Future Shock* appeared, the widespread sense of constant, rapid change has only deepened. But rather than being shocked by it, we now regard acceleration as a central part of modern life. Everyone assumes that each passing year will bring faster, cheaper, sleeker, and more powerful technologies. Not a week goes by without headlines about new breakthroughs in artificial intelligence, biomedical research, and other promising vistas of progress. In many ways, we have become economically and culturally dependent on technological progress, and futurists have assumed the role of scouts who will help us navigate the misty terrain in front of us.

Chasing Plausibility

The most valuable recent contributions from futurists has been in the field of disaster preparation and risk management. In 2008, Jane McGonigal of the (nonprofit) Institute for the Future in Palo Alto led an exercise in which thousands of people worldwide imagined that it was 2019 and they were living through a pandemic of a novel respiratory virus. Then, in 2010, she ran another simulation game, set in 2020, where "nearly twenty thousand players showed up to predict what actions they could take to help others during a complex outbreak of possible future global crises, including a pandemic and extreme weather from climate change happening at the same time" (McGonigal 2022, p. xv).

The 2010 exercise imagined that a flu-like pandemic would emerge out of China, that misinformation would frustrate the response, that wildfires would sweep the US West Coast, and that millions would suffer power outages at a time when they were supposed to be remaining in place. Reflecting back on this simulation in her 2022 book, *Imaginable: How to See the Future Coming and Feel Ready for*

Anything—Even Things That Seem Impossible Today, McGonigal writes, "The storylines that we wrote a decade in advance turned out to be pretty much exactly what we saw in the headlines of the real 2020 and 2021" (McGonigal 2022, p. xx). COVID-19 had been accompanied by West Coast wildfires, conspiracy theories, and then widespread power outages in Texas.

McGonigal thinks everyone should train their minds to think more like a futurist, imagining scenarios a decade in the future that seem impossible today. "The purpose of looking ten years ahead isn't to see that everything will happen on that timeline," she writes, "but there is ample evidence that almost anything *could* happen on that timeline" (McGonigal 2022, p. 8). By drawing on "past experiences, current hopes and fears, and your intuitions about what might change in the years ahead," she explains, you can "bring something into the world that doesn't exist yet. ... making a brand-new memory of something you haven't even lived through yet." And once that happens, McGonigal observes, "what was previously unimaginable to your brain is now *imaginable*" (McGonigal 2022, p. 26).

However, while it is prudent to consider underappreciated or underestimated risks that may lie ahead, futurist thinking—as this book will show—also encourages an overestimation of the likelihood of positive breakthroughs and possibilities. As McGonigal herself notes, an ample body of research in psychology finds that "imagining a possible event in vivid, realistic detail convinces us that the event is more likely to actually happen" (McGonigal 2022, p. 140). The futurist methodology rests on a foundation of radical open-mindedness, even willful gullibility.

In coaching her readers to play around with future scenarios, McGonigal's first rule is to suspend disbelief: "In order to train your brain to think the unthinkable, you have to actively pit possibilities in your brain *that your brain will naturally resist*. You don't want something to be unimaginable simply because you refuse to imagine it. Be willing to think hard about ideas you would normally dismiss as impossible, impractical, or even dangerous" (McGonigal 2022, p. 44).

This approach is captured by "Dator's Law" (coined by the futurist Jim Dator), a fundamental principle of futurist methodology according to which "Any useful statement about the future should at first seem ridiculous." Having embraced this ethos, McGonigal considers it a "point of pride" that she is "very rarely stumped with a truly unchangeable fact" about the human experience. If challenged with a statement like "It takes a man and a woman to make a baby," she will go in search of ways to prove the assertion wrong (in this case by citing experimental fertility treatments that combine "the genetic material of *two women* and one man to make one baby") (McGonigal 2022, p. 67).

Another such statement, "the sun rises in the east and sets in the west every day," leads McGonigal in search of signs that humans might soon travel to Mars, where sunrises and sunsets "don't happen every day – at least, not by our standard definition of a 'day' on Earth." As "evidence" of this possibility, she cites the fact that "there are plenty of space entrepreneurs trying to develop the technology to help humans settle on Mars as soon as possible" (McGonigal 2022, p. 72). But as we will see in Chap. 5, the claims made by entrepreneurs promising to send humans to Mars aren't really evidence at all.

Indeed, such figures have a financial interest in creating the impression that exceedingly difficult feats are eminently plausible and within reach. It is little wonder that the futurist discipline and the tech industry are so closely intertwined. All are in the business of selling the future. While a well-meaning educator like McGonigal wants us all to be "ready to believe that almost anything can be different in the future," there are many more in Silicon Valley who stand to gain from having a public that is ready to believe anything (McGonigal 2022, p. 194).

References

Bell, Wendell. 2017. *Foundations of Future Studies*. Vol. 1. New York: Routledge.

Chesterton, G.K. 1904. *The Napoleon of Notting Hill*. London: John Lane.

Clarke, I.F. 1979. *The Pattern of Expectation, 1644–2001*. New York: Basic Books.

Condorcet, Antoine-Nicholas Condorcet. 1795/2009. *Outlines of an Historical View of the Progress of the Human Mind*. Chicago: G. Langer.

Darwin, Charles. 1859. *On the Origin of Species*. Project Gutenberg. https://www.gutenberg.org/files/1228/1228-h/1228-h.htm.

de Jouvenel, Bertrand. 1967. *The Art of conjecture*. New York: Basic Books.

Dickson, Paul. 1971. *Think Tanks*. New York: Ballantine Books.

Future Today Institute (FTI). 2022a. *Methodology & Frameworks* (Volume 01). In 2022 Tech Trends Report, 15 Edition.

———. 2022b. *Artificial Intelligence* (Volume 01). In 2022 Tech Trends Report, 15 Edition.

———. 2022c. *Logistics, Robotics & Transportation* (Volume 09). In 2022 Tech Trends Report, 15 Edition.

———. 2022d. *Synthetic Biology, Biotechnology & AgTech* (Volume 12). In 2022 Tech Trends Report, 15 Edition.

Greely, Henry T. 2019. CRISPR'd Babies: Human Germline Genome Editing in the 'He Jiankui affair'. *Journal of Law and the Biosciences* 6 (1): 111–183.

Luckhurst, Roger. 2017. Introduction. In *The Time Machine*, by H.G. Wells, vii-xxv. Oxford: Oxford University Press.

McGonigal, Jane. 2022. *Imaginable: How to See the Future Coming and Feel Ready for Anything – Even Things That Seem Impossible Today*. New York: Spiegel & Grau.

Segal, Howard P. 2006. *Technology and Utopia*. Washington: American Historical Association.

Shelley, Mary. 1992. *Frankenstein*. New York: Alfred A. Knopf.

Toffler, Alvin. 1970. *Future shock*. New York: Random House.

Wager, W. Warren. 1983. H.G. Wells and the Genesis of Future Studies. World Network of Religious Futurists. http://www.wnrf.org/cms/hgwells.shtml. Accessed 27 July 2022.

Webb, Amy. 2022. *Should We Use Gene Editing to Make Better Babies?*. Intelligence Squared US. https://www.intelligencesquaredus.org/debate/use-gene-editing-make-better-babies. Accessed 8 Aug 2022.

Wells, H.G. 1902. *Anticipations of the Reaction of Mechanical and Scientific Progress Upon Human Life and Thought*. 2nd ed. London: Chapman & Hall, LD.

———. 1913. *The Discovery of the Future*. New York: B.W. Huebsch.

Chapter 3
The Horizon Bias

Contents

Our capacity to engage with civilizational problems is handicapped by the horizon bias. This psychological tendency leads us to systematically underestimate the challenge of solving civilizational problems with technology. Indeed, it is both easy and exciting to imagine technological solutions appearing out of nowhere. When we declare metaphorical war on a problem like cancer or climate change, the implicit hope is that a *deus ex machina* will arrive to deliver victory just in the nick of time. Though we know we should not bet the farm on such expectations, it is all too tempting to envision solutions that would make problems like climate change, pandemics, and cancer just *go away*.

But such pining can hamper our ability to prepare for an intrinsically uncertain future. Proper preparation demands that we not rely on a grievously biased sample of past experiences. As we confront civilizational problems, we must avoid acting like gamblers who retain vivid memories of those rare occasions when they hit it big while forgetting the more numerous occasions when the machine swallowed their money.

The impulse to declare war on civilizational problems is born of the same tendency. When we call for a wartime mobilization against the coronavirus or climate change, we are thinking of World War II. Its hideous death toll notwithstanding, WWII is remembered as what Studs Terkel called the Good War. To observers from the third decade of the twenty-first century, it offers an appealingly straightforward moral: it was the occasion when the forces of freedom and democracy came together to defeat fascism and genocide.

But, of course, the Good War has turned out to be a rather unhelpful paradigm for military conflicts since then. The good-versus-evil framing didn't work out so well in Vietnam, nor has it achieved the desired result against what former US President George W. Bush called the "Axis of evil": Iran, North Korea, and Saddam Hussein's Iraq.

We are similarly selective in what we learn from the history of technological progress. Everyone is familiar with the sequence of events that followed from US President John F. Kennedy's May 25, 1961, speech to Congress announcing his plans to send Americans to the moon (Kennedy 1961). That mission was indeed accomplished. But we should not forget all the occasions when hoped-for moonshots have not led to man-on-the-moon moments.

Such dashed hopes have been a recurring feature of the War on Cancer. After a half-century of repeated failures to come up with "the cure," we should be skeptical of those who still insist that this outcome is right around the corner. In February 2021, the Harvard aging researcher and best-selling author David Sinclair told the comedian-hosts of the *Smartless* podcast that, "This isn't rocket science. It's actually, it's going to be easy to cure aging and cancer" (Sinclair 2021).

Ironically, Sinclair is correct that finding the cure for cancer "isn't rocket science." If it was, we would have already done it. In issuing such a confident prediction, he—like many others—is capitalizing on our persistent hope that in an age of rapid technological progress, the decisive breakthrough surely must be imminent. Nowadays, we eagerly expect "the cure" to come from immune-oncology, because we have already forgotten that we had the same level of anticipation when it came to anti-angiogenesis drugs in the late 1990s, and to many other previously hyped interventions.

The horizon bias afflicts us all, but it is most consequential in those who should know better; that is, in those with enough expertise to be able to offer scientific/technological solutions to civilizational problems in the first place. As a problem among the scientifically informed, the horizon bias is born not of ignorance but of forgetfulness. It is a lack of historical awareness—a failure of memory—that leads the scientifically well-informed to offer overly confident statements about the future.

Generally, these historical omissions are not really omissions of bare fact. Cancer researchers are well aware of the War on Cancer's timeline. They know, for example, about Judah Folkman's discovery of anti-angiogenesis drugs like Avastin (bevacizumab), which they may sometimes prescribe to some of their sickest patients. But while they can recount the bare facts about anti-angiogenesis therapies, they usually have forgotten (or are too young to know) how the arrival of these therapies made people at the time feel about the future.

The failure here is thus *affective* rather than factual. The same feelings of confidence and excitement surrounding current research were felt by those engaged in earlier research that ultimately did not pan out as hoped. A gambler who remembers all his losses as vividly as his wins presumably would not spend as much time pumping more quarters into the machine. By the same token, if we can recall all the previous occasions when our excitement around technological advances ultimately turned into disappointment, we might be more open to exploring pathways and solutions that do not depend solely on such breakthroughs.

The Problem of Knowing Too Much

Why might technologists, scientists, and other experts systematically overestimate humanity's ability to end pandemics, solve climate change, win the War on Cancer, or colonize Mars? All too often, the expert prognosticator fails to distinguish between what we call *manifesto science* and *normal science*. As we will see, the horizon bias is manifest in manifesto science. Precisely because they have less knowledge, laypeople are less inclined to conflate these two varieties of scientific thinking—the first of which concerns a destination over the horizon, and the second of which involves the step-by-step journey needed to get there.

Whether the cause is ego or just eagerness, those with deep firsthand knowledge in the hard sciences can fall into the trap of thinking they are "Laplace's Demon." In a famous passage in his 1814 *Philosophical Essay on Probabilities*, the French mathematician and philosopher Pierre-Simon Laplace asked his readers to imagine what it would mean to have absolute knowledge in a universe that is subject to fundamental laws of motion. Such a being would be able to foresee every movement and every interaction of every atom, and nothing about the future would be uncertain or unknown (Laplace 1814, p. 12).

Laplace's hard determinism was more or less vanquished a century later by the development of quantum physics, which introduced randomness (and thus uncertainty) into the cosmic equation. Yet if the universe really was structured in the way that Laplace had assumed it to be—like billiard balls bouncing off one another but always following Newton's laws of motion—an omniscient intelligence actually could, in principle, foresee every future movement and every transference of energy from here to eternity. How was Laplace to know that there was a deeper level of reality that followed different rules?

This chapter will show that Laplacian demonic possession all too often appears in experts' predictions about the trajectory of their own research. Like Laplace and his hypothetical intelligence (he himself didn't use the term "demon"), those who know more than everyone else about a topic sometimes convince themselves that they know *enough* to anticipate how future events will unfold.

As the psychologist Gerd Gigerenzer has shown, predictions can be derailed because they are based on insufficient information, but also because they are *too* informed. In the face of an uncertain world, Gigerenzer points out, only some information is useful for anticipating the future, whereas many other data may merely be clouding the picture. The key is not necessarily to have more data but to intuit which data matter (Gigerenzer 2019). This problem has become especially acute in the age of data mining and other efforts to understand highly complex probabilistic systems. Just because you can amass more and more data from which to draw inferences doesn't mean that you've amassed the *right* data for the problem at hand.

Cancer researchers have been especially susceptible to these mistakes. (They have also had more opportunities to make them, because the public is constantly demanding confident predictions and new breakthroughs in the War on Cancer.) Though there has been notable progress against some cancers since 1971, that isn't really our focus. The child in the backseat of the car doesn't ask, "How far have we traveled?" She asks, "Are we there yet?"

As the ones behind the wheel, cancer researchers are tempted to believe that they know the answer. And because they are focused primarily on the scientific facts, they can see how every scientific advance against cancer seems different from the ones that came before. To the scientifically informed, the details about anti-angiogenesis drugs that fueled hopes for "the cure" in the late 1990s look very different from the facts about the immunological approaches that are fueling hopes today.

In reality, no one can know even if we've reached the halfway point of our journey. But with all the knowledge and treatments that we have developed already, we want to believe that the end is just around the corner. Moreover, prosecuting the war requires that we have an end in sight. Science cannot be conducted as if it were a joy ride. Scientists need clearly articulated goals, some of which will be modest—concerning the potential outcome of a single experiment—and some of which will be much more ambitious, such as curing cancer or colonizing Mars. Without such objectives, there would be no way to organize or direct one's research.

But it is one thing to confidently set a difficult goal; it is quite another thing to believe that a miracle breakthrough is nigh. Too much confidence can lead to recklessness and create a moral hazard. Why worry about carbon emissions if we can anticipate that DAC or some kind of carbon-eating nanobot will eventually be deployed to reverse climate change? If the forthcoming technological fixes really are as powerful as advocates of technological progress would have us believe, we should enjoy the economic benefits of burning coal and petroleum for as long as we can.

Manifesto Science vs. Normal Science

Carbon-eating nanobots and other provocatively ambitious technological aspirations are the stuff of manifesto science, which is not a bad thing in itself. Manifesto science should be understood as something related to, but separate from, what the philosopher of science Thomas Kuhn called "normal science." By that, he meant research that builds on previous discoveries that have been generally accepted by a community of scientists working in the areas and on the same kinds of problems (Kuhn 1962, p. 10).

Normal science seeks to add to what is already known in a given domain. A molecular biologist might offer a testable hypothesis about, say, protein folding, and then design an experiment capable of falsifying or lending further support to his conjecture. Normal science, then, *is when we set goals that can feasibly be achieved by completing a single experiment, or a related set of experiments.*

By contrast, a more ambitious project would be to make viral pandemics impossible by producing vaccines that offer perfect protection without any side-effects. The key difference, here, is that this undertaking cannot be completed with one experiment; it necessarily would involve many iterations of normal science. That fact alone changes everything, because it means that we are engaged in manifesto science, the objective of which is not immediately accessible. *We have established a goal whose achievement will depend on our success in achieving many smaller victories through the course of normal science.*

Because it depends on meeting certain conditions in sequence, manifesto science follows the logic of the R.E.M song: "Can't Get There From Here." We can guess at what hypotheses we will be testing at some later stage in the journey, but we can't actually know what it will be until we're there; each step of normal science along the way may force us to take a detour. The outcome of each of subsequent experiment will determine the hypothesis and design of the experiments that follow.

We are engaged in manifesto science when we look to the horizon, name our destination, and chart a general course forward. As soon as we come to a river or fjord, we will be tasked with performing normal science. We will have to wade in, not knowing how deep the water is, nor how many more hazards await us on future legs of the journey.

Suppose that you are testing a new potential treatment for leukemia. You know from experience not to expect a new "wonder drug" or "magic bullet," and you are fully prepared for a scenario in which the drug fails. Still, you are confident in the compound's "bioplausibility," so you conduct a clinical trial to determine if it offers "meaningful benefits" above and beyond the current standard of care.

If the test group fares better than the control group, you can accept the new compound as an effective treatment against leukemia. But, of course, the more likely outcome is that the drug will fail. Between January 2000 and October 31, 2015, just 3.4% of investigational cancer treatments made it from phase 1 clinical trials through to FDA approval (Wong et al. 2019, p. 273).

How-Possibly Explanations

Given the difficulties of making accurate forecasts about normal science, we should expect even less from manifesto science as a window onto the future. It is a standard principle of forecasting that the further ahead one looks, the harder it is to make reliable predictions. But beyond that problem, there are also difficulties associated with the relationship between normal and manifesto science.

As the source of ambitious, expansive goals, manifesto science is a powerful motivator, mobilizing not just individual researchers but entire scientific communities. It also tends to be closely bound up with the sources of necessary funding, be it from public institutions (as with the Human Genome Project) or billionaires like Elon Musk and Jeff Bezos, both of whom harbor plans to settle humans on Mars.

At its best, the manifesto perspective advances what the twentieth-century philosopher William Dray called "how-possibly science," which focuses on feasibility rather than actuality. For example, Dray asked his readers to consider the "following extract from the "Parade" column of a popular magazine:

> An announcer broadcasting a baseball game from Victoria, B.C., said: 'It's a long fly ball to centre field, and it's going to hit high up on the fence. The centre fielder's back, he's under it, he's caught it, and the batter is out.' Listeners who knew the fence was twenty feet high couldn't figure out how the fielder caught the ball. Spectators could have given them the unlikely explanation. At the rear of centre field was a high platform for the scorekeeper. The centre fielder ran up the ladder and caught the ball twenty feet above the ground. (Dray 1957, pp. 390–391)

Philosophers of science regard how-possibly explanations as an indispensable element of science, because, as David B. Resnik argues, they provide benchmarks for progress, as well as theoretical roadmaps for pursuing further investigations when empirical data is lacking (Resnick 1991). Indeed, how-possibly explanations may be a necessary (though not sufficient) prerequisite for arriving upon good "how-actually explanations."

If You WILT It, It Is No Dream

How-possibly explanations are also part and parcel to technological innovation—and to techno-utopian hype. Consider the work of the biomedical gerontologist Aubrey de Grey, whose proposed solution to cancer goes by the acronym WILT: whole-body interdiction of lengthening of telomeres.

De Grey has achieved fame (or notoriety) for his plan to extend human life spans to the range of centuries or even millennia. To that end, he offers his signature Strategies for Engineered Negligible Senescence, which promises to prevent or repair all the damage that accumulates in human bodies as we age.

Cancer is a significant obstacle for anyone who wants to live to be a thousand. Not only does cancer become more common as we age, because there has been more time for cell damage and dangerous mutations to occur; it also becomes more *likely*, because an older body has a less efficient immune system and a larger collection of senescent (retired) cells that create a more toxic microenvironment (Raza 2019, p. 81). Thus, even if someone can evade the depredations of heart disease, dementia/Alzheimer's, and other big killers, cancer eventually will emerge as the dark handmaiden of aging.

When it does, one will be subjected to the standard protocols of surgery, chemotherapy, or radiation therapy, but one won't know whether one has been "cured." The doctor may report the happy news that there is "no evidence of disease" (NED), but that won't really tell you whether the remission will be "deep" or short-lived. If you die 10 years later from a heart attack, your relatives will probably look back on your cancer and say it had been cured. If you survive the heart attack but die 2 years later from a resurgence of your cancer, they will conclude the opposite.

The "cure" will be a matter of retrospective reflection. Doctors cannot reliably rid a cancer patient's body of every single cancerous cell. But if they can provide treatments that generally will prevent cancers from cutting our lives short, we might as a society conclude that we have achieved a certain therapeutic ideal in the war against the disease.

De Grey, however, would disagree. Because his millennial ambitions have set the bar so much higher, cancer becomes an even bigger problem. The prostate cancer that goes untreated because it is growing so slowly that one will die *with* it but not *because of* it suddenly becomes an issue that cannot be ignored. Once you've removed all the barriers to a 100-year lifespan, you will then confront new barriers on the way to 200, then 300, and so on. The prostate cancer that wasn't really a problem at 90 might not kill you until you're 150. Most would consider living to that age to be an impressive achievement. But for someone who aspires to live to a thousand, it could only be regarded as a tragic disappointment (De Grey and Rae 2007).

The solution, according to de Grey, is to develop a "super" cure that would make cancer impossible, by removing from each cell of our bodies the gene that produces the enzyme telomerase. With each cell division, the telomeres that protect that cell's DNA grow shorter. Telomerase has the capacity to restore telomeres, but the gene that produces it is switched off in the vast majority of cells.

The evolutionary reason for this, most likely, is to protect against cancer, by stamping cells with a finite lifespan. Thanks to the erosion of telomeres, some cancerous cells that would otherwise kill us reach the end of their leashes. And yet, many other cancerous cells will break the leash by switching on the telomerase gene. That's where de Grey comes in. If we remove the telomerase gene from every cell, no cell could exceed its preprogrammed lifespan.

But purging the telomerase gene would not be sufficient, not least because some cancers spread by means other than telomerase. Moreover, it isn't even an option to excise the telomerase gene in the cells that are supposed to have it switched on, including those that make platelets and red and white blood cells, as well as gut and lung tissue, and skin. Ridding these cells of telomerase would promptly kill us. To get around this problem, de Grey suggests that we use stem-cell technologies to replace these cells as they die off. The cancer-free Methuselahs in his imagined future would undergo regular infusions of stem cells with long telomeres but no telomerase genes.

This is manifesto science. De Grey's anti-aging project describes a solution to cancer based on yet-to-be falsified conjectures that he assumes will turn out to be true, and depending on a sequence of experiments, many of which cannot be performed yet.

Miracle Thinking

Because it is concerned with distant goals, manifesto science—and those engaged in it—is highly vulnerable to the horizon bias. The key to distinguishing good from bad manifesto science is to interrogate the implicit how-possibly explanations that are in play. Consider the following two ambitious goals:

Goal 1: To deepen humanity's understanding of the causes of the late-Cretaceous extinctions by designing time-travel technologies that will retrieve specimens from that epoch.

Goal 2: To solve the problem of climate change by designing nanobots that can consume atmospheric carbon-dioxide.

When scientists practice manifesto science, they should avoid the error depicted in a famous Sydney Harris cartoon. Two academics are standing in front of a black-board that purports to show a mathematical proof, except that at one stage of the process, it states: "And then a miracle occurs." In what one imagines to be a deadpan tone, the second academic says to the first: "I think you should be more explicit here in step two."

Good manifesto science limits itself to a sequence of steps that we have reason to believe are feasible given what we currently know. One cannot simply assume that a scientific breakthrough will supply a critical missing element at some point in the future. This is the problem with Goal 1, which would appear to require meta-physical impossibilities. Until we have established that time travel would not violate any fundamental law of nature, Goal 1 will never generate good manifesto science.

However, there are ambitious goals that would make for bad manifesto science at one time, but good manifesto science at a later date. History is replete with exam-ples of people lacking how-possibly explanations for technological achievements that later turned out to be possible. Timing is everything. Many mundane technolo-gies of the twenty-first century would rightly be regarded as miracles in the seven-teenth century, not just because they would be unfamiliar to people from that time, but because they would have been impossible at that time.

Around 1620, the Anglican bishop Francis Godwin published *The Man in the Moone* (which is thought to have been inspired by the Copernican Revolution), wherein a Spaniard named Domingo Gonsales discovers a species of wild swan that is capable of transporting him to the lunar surface. Within 350 years, humanity had made this fantasy a reality, because it had come to understand that traveling to the moon was well within the bounds of physical possibility.

But neither Godwin nor any of his contemporaries could have known that, nor could they have offered credible how-possibly explanations for space travel based on the technologies and scientific knowledge of their own time. If we put ourselves in their shoes, we might wonder if some future generation will look back on our own time as a quaint period before the advent of time travel.

Still, we also would have to recognize Godwin's fictional account as a case of bad manifesto science. The fact that later generations did indeed go to the moon does not retroactively validate the scientific merits of Godwin's story. If pressed for more scientific details about how his swan would complete the journey through space, he inevitably would have had to resort to: "And then a miracle occurs." He could argue that, "For all we know, in 350 years, humans will discover the means to transport ourselves to the moon"; but such speculation tells us nothing. Unless the manifesto science can describe all the steps of normal science necessary to achieve a stated end, it is no better than metaphysical wishful thinking.

There is also an ethical dimension to manifesto science. Good manifesto science can and should establish goals that are worthy of pursuit by scientific communities working with finite resources. Insofar as bad manifesto science encourages us to spend money on ambitious goals that are unachievable—either in principle or at this point in time—or simply inferior when weighed against other priorities, it should be rejected.

So, where does WILT stand? On the one hand, it does not rely on unexplained magical interventions, which means that it could be good manifesto science. De Grey has offered enough how-possibly explanations to set himself apart from those who would ask us to fund research into time travel (De Grey and Rae 2007). By establishing concrete goals for researchers to pursue through normal science, WILT distinguishes itself from Godwin's swan. On the other hand, de Grey errs when he steps into the realm of prediction. He has offered a plausible destination and road-map, but he is far too confident that there will be no major obstacles along the way.

Manifesto science cannot escape the "can't-get-there-from-here" problem. Suppose we decide that de Grey has spelled out WILT to a degree that would qualify as good manifesto science, meaning there are no ineliminable steps in which "a miracle occurs." Suppose, also, that experiments will validate a specific application of gene editing that can remove the telomerase gene from every cell in the human body. Even then, we can't know whether we will succeed in introducing stem cells with long telomeres so as to preserve the life of the telomerase-free human we've created.

The point is that manifesto science is not only sequential, but *obligatorily* so. It involves more than one iteration of normal science, and it sets goals that will only become accessible after a number of other conditions have first been met. Manifesto science thus has a depth that sets it apart categorically from normal science; it also relies wholly on normal science proceeding in a "just-so" fashion.

Put another way, manifesto science is an odyssey. When Odysseus departs from Troy, he knows the general direction in which he needs to sail, even though he cannot actually see Ithaca within his visible horizon. But he cannot possibly foresee all the additional steps that the gods will require of him. Even when he leaves Circe's island with the goddess's wisdom and blessing, he first must travel to the borders of the underworld and then back to the island to bury a fallen comrade before even beginning to set sail again for home—a series of obligatorily sequential steps.

The more individual steps of normal science that are required, the harder it becomes to predict the final outcome. The odds of ten specific conditions being met are lower than those of just two or three conditions being met. When we imagine achieving the ends proposed by a manifesto science of cancer, we can tell ourselves how-possibly each of the steps of normal science will work, and that sense of knowing will give us confidence. What if each step of normal science were to go as planned? We would reach the final destination right on time.

There is a certain convenience to this mode of thinking. Because we can only speculate about the later stages of the obligatory sequence, we have a license to elide the messy contingencies that are inevitable in normal science. Succumbing to the horizon bias, we can say things like, "All we would need to do is provide regular infusions of stem cells with long telomeres, but without telomerase genes." Precisely because we don't yet know what steps of normal science this would require, we can imagine it as eminently doable.

Moreover, because we are so eager for progress, the uncertainty lends itself not to doubt but to hype. We are deeply encouraged by the fact that the objective is not logically impossible, nor obviously in breach of the laws of physics. We have faith in progress, trusting that anything abiding by the laws of physics is possible with the correct information (Deutsch 2011, p. 56). We forget that inductive reasoning based on past experience advises caution. But if we remember that unforeseeable obstacles will reliably present themselves through the course of normal science, we will be more careful about assuming that a point visible on the horizon is within striking distance—or that it even represents the final destination we are seeking.

The horizon bias does not imply that technological solutions to civilizational problems won't happen soon. It is certainly possible that some lone genius could crack the problem of cancer or climate change tomorrow. Working out of her garage, perhaps she has come up with a cheap and supremely efficient way to sequester atmospheric carbon at scale, allowing us to return to pre-industrial levels within a year.

In this scenario, pessimistic claims about the future will have been falsified, but claims about our rational expectations will not have been. If you announce that you've just bought a lottery ticket and simultaneously made a bid on a mansion that you can't afford, no one will commend you for your financial judgement, even if you win.

Moonshots versus Mars-shots

To see how the horizon bias distorts our expectations of highly ambitious technological "moonshots," we can revisit the original source of that favored metaphor. Just as World War II continues to inspire wars on big problems, the first actual moonshot remains a favored metaphor for framing ambitious plans with the potential to yield enormous returns.

The moon landing holds canonical status for good reason: it confirmed an accurate truly ambitious technological forecast. On May 25, 1961, US President John F. Kennedy told Congress, "I believe this nation should commit itself to achieving the goal, before this decade is out, of landing a man on the moon and returning him safely to Earth" (Kennedy 1961).

Though he didn't live to see it, that goal was achieved, inspiring Nixon and his advisers to set a deadline of just 5 years for discovering the cure for cancer. Nowadays, many are looking at the new mRNA vaccines and wondering if they could be a platform for launching a moonshot to solve the problem of infectious disease entirely. Others, meanwhile, see similar potential in direct-air capture as a possible solution to climate change. And, of course, new-generation space entrepreneurs like Elon Musk seem to be channeling Kennedy directly when they suggest that we might be sending humans to Mars by as soon as 2026. But there is a big gap between Kennedy's moonshot and Musk's proposed Mars-shot.

One difference lies in the number of obligatorily sequential and contingent steps needed to reach the destination. Remember, expressions of manifesto science can be distinguished from one another by their depth. While the Apollo program depended on many successful applications of normal science, much of this work could be pursued simultaneously. NASA scientists did not first have to make Step A work before they could even begin to think about Step B; rather, one team could work on Step A and another on Step B at the same time. Those tasked with building the lunar module did not have to wait for the booster rockets to be finished before they could even begin their work. Though the project required substantial sums of money, manpower, and know-how, it was ultimately a much "shallower" challenge.

The moonshot team was pursuing an objective that was already visible within the horizon of normal science. Each of the individual objectives needed to make the overall project succeed qualified as what philosophers of science call an "accessible goal": they could be reached "in a finite number of steps in a finite time" (Niiniluoto 2019).

As soon as one starts looking to the horizon, the task becomes more complicated. The horizon marks the point beyond which one cannot see. Though it might seem as though a vigorous day's travel can take you to that point, past experience teaches us that this never occurs. The horizon bias clouds our vision of future technological developments. Each day of the journey reveals things that had lay beyond an earlier horizon.

Musk is certainly not averse to making exciting predictions. His ability to generate hype is surely a factor in his ascent to becoming one of the world's richest men. If anyone else were to say that there will be a city of one million people on Mars by 2050, we would probably ignore them. But when Musk makes this claim, he is tapping into his reputation as the genius behind Tesla and SpaceX. Nobody wants to be like the investor who lost his shirt by shorting those successful endeavors early on.

Musk's plan for building and populating self-sustaining cities on Mars is a case of manifesto science. But it implies a process much deeper than the one needed to land a man on the moon. The journey from our current understanding to a future in which we are ferrying colonists to Mars is long and obligatorily sequential. Someone who placed a successful wager on Kennedy's forecast would be unwise to accept the same odds for Musk's. The long sequence of iterations of normal science needed to establish Martian cities implies that there will be many unforeseen failures and delays along the way.

Likewise, our ability to defang COVID-19 with a safe and effective vaccine does not imply that we will have the same success against cancer. Although the low success rate of investigational cancer treatments should militate against optimism, we are nonetheless prone to seize on each rare "breakthrough" as representing an "inflection point" (another common and ultimately misleading metaphor that we will examine in more depth in a later chapter).

A major example in recent decades is the Human Genome Project, which was officially launched in 1990, to sequence and map all the genes that constitute *Homo sapiens*. In June 2000, President Bill Clinton gave a speech to mark the publication of the "working draft" of the human genome. "It is now conceivable," he averred, "that our children's children will know the term cancer only as a constellation of stars" (Clinton 2000). Sadly, Clinton now has grandchildren, and they will almost certainly grow up to know cancer as the name of a terrible disease.

For its part, the medical and scientific community saw far-reaching potential in the burgeoning field of genomics. By revealing the genetic bases of disease, researchers hoped that it would be only a matter of time before targeted gene therapies were developed. Owing to the work of biologist Robert Weinberg and others, cancer, too, had come to be understood as a genetic disease, and so it was hypothesized that the cure might lie in repairing genetic errors, some of which had already been identified.

Moreover, at the turn of the century, we knew about oncogenes and proto-oncogenes, and we knew that mutations in tumor-suppressor genes could shut down or bypass the body's defenses against cancer cell growth. The working draft of the human genome thus promised a future in which all these genetic factors in cancer had been identified and subjected to genetic-modification technologies. Human genomes would be constantly surveilled, and any glitches would be automatically fixed by powerful new genetic-repair technologies. We would all have our own biological malware-removal software. It is yet to happen.

Unlike Kennedy, Clinton was merely expressing hope, not making a forecast. Yet it was Clinton who fell for the horizon bias. He presented an exciting possibility as conceivable, while avoiding the epistemic obligations associated with offering a prediction. Ostensible "breakthrough" events like the sequencing of the human genome and the creation of mRNA vaccines are teachable moments. On such occasions, one can make two apparently contradictory observations about the future.

On the one hand, a relevant scientific discovery provides a legitimate occasion on which to express hope. Clinton's line about his grandchildren knowing cancer "only as the name of a constellation of stars" was not necessarily inappropriate at the time. But on the other hand, our long history of underwhelming progress against cancer made it predictable that Clinton's high hopes would be dashed. After all, those high hopes were, in part, a product of ignorance. In the year 2000, much of the normal science on how to modify cancer genes still needed to be done. Interventions that looked effortlessly straightforward in 2000 would predictably look very difficult by 2020.

The Limits of Epistemic Possibility

The excitement over genomic sequencing was born of epistemic possibility. If you are in a building during an earthquake, you should treat that building's collapse as an epistemic possibility. *For all you know*, the building could collapse. Though it may have special structural features rendering it quake-proof, you don't know about those. Put more broadly, the domain of epistemic possibility expands with ignorance. The less one knows, the more one can suppose to be possible.

But the excitement over the genomic sequencing should not be lumped together with the excitement that some feel toward, say, homeopathic treatments for cancer. The latter can be dismissed out of hand, because homeopathy is sheer quackery, based on an ignorance of both the disease and the purported cure. The difference concerns what philosopher Keith DeRose calls "relevant communities" (DeRose 1991, p. 584). In the case of cancer research, the relevant community refers to those best informed about cancer biology. The advent of genomics was exciting to those with the deepest knowledge of cancer biology precisely because it promised things that even they with all their knowledge could not rule out.

The result is an epistemic paradox. While a deeper knowledge of the relevant science allows one to access a wider range of how-possibly explanations, the incompleteness of that knowledge widens the scope of epistemic possibility. At such breakthrough moments, the expert knows both too much and too little at the same time. It is on such occasions that the relevant community looks to the horizon and adopts or revises its manifesto. When the best informed can sincerely say, "*For all we know*, this latest advance could swiftly lead to highly effective cancer treatments," that is a legitimate cause for excitement.

Regardless of what Clinton knew, we can assume that his science advisers were not ignorant about cancer and genetics. In their view, the next step after sequencing the human genome would be to determine how existing genetic-modification techniques could be used to correct the mutations that cause cancer. It was epistemically possible that these genetic fixes would be quite straightforward. *For all they knew* at the time, there might be only a few ways that proto-oncogenes could transform into oncogenes, or that tumor-suppressor genes could become non-functional. If that had been the case, genetic engineering might well have become the long-awaited silver bullet against cancer.

One scientist who made the same mistake as Clinton was the Nobel laureate biologist James D. Watson. In the spring of 1998, Watson was reported in the *New York Times* to have predicted that the tumor researcher Judah Folkman was "going to cure cancer in 2 years," following his success with anti-angiogenesis drugs that prevented tumors in mice from securing nourishment by hijacking local blood vessels. According to the reporter, Gina Kolata, "Within a year, if all goes well, the first cancer patient will be injected with two new drugs that can eradicate any type of cancer, with no obvious side effects and no drug resistance – in mice" (Kolata 1998). Since she needed to translate these laboratory findings into terms that would resonate with the lay public, Kolota emphasized the against-all-odds nature of

Folkman's discovery. Researchers cautioned that "the history of cancer treatments is full of high expectations followed by dashed hopes when drugs with remarkable effects in animals are tested in people," she wrote. "Still, the National Cancer Institute has made the [new anti-angiogenesis] drugs its top priority, said Dr. Richard D. Klausner, the director." Folkman's discovery, Klausner added, was "the single most exciting thing on the horizon" (Kolata 1998).

Over two decades later, Watson's failed forecast has become the stuff of legend. Although he later denied ever having said such a thing, there is no evidence that he sought to deny Kolata's story when it first came out. Indeed, according to the distinguished cancer researcher Vincent DeVita, Watson doubled down on his original forecast. In his memoir, DeVita recounts a meeting in September 1999 where Watson declared while he had been wrong about Folkman curing cancer 2 years earlier, he believed that he would still do it in the coming 2 years (DeVita and DeVita-Raeburn 2016). There is no reporting on Watson's thoughts 2 years thereafter; but it is safe to say that in September of 2001, the world's attention was directed elsewhere.

In any case, we should recognize that Watson perhaps had little to lose by making such a confident prediction. Forecasts tend to follow the same logic as Kennedy's quip about the Bay of Pigs fiasco: "Victory has a thousand fathers, but defeat is an orphan." Moreover, we can take Watson's pronouncement as evidence of the excitement that attended Folkman's work. It is this affective side of the story that is so often forgotten, thereby distorting our expectations for each new promised breakthrough.

Additional evidence of the since-forgotten mood at the time is the movement in markets. In 1998, EntreMed, the company with the licensing rights to Folkman's two compounds, experienced a fourfold increase in its stock price within the space of just a few days. As economists Gur Huberman and Tomer Regev pointed out, the price movement was driven wholly by "contagious speculation" over a "nonevent" (Huberman and Regev 2001, p. 387).

By "nonevent," they meant Kolata's story and its quotes containing overwrought predictions from Nobel laureates, not Folkman's discovery, which had already been reported 5 months earlier in *Nature*, and 6 months earlier by another reporter from the *New York Times*. In a November 1997 story, Nicholas Wade offered a comparatively more circumspect account of Folkman's research, eschewing hifalutin praise from famous scientists. The market reaction on that occasion was duly muted (Wade 1997).

What does this episode teach us? For one, we should be wary of confident predictions regardless of who makes them. Just because someone has a Nobel prize in a related scientific discipline doesn't mean they have any special insights into the future of cancer research.

Moreover, it should be noted that Folkman's work did at least lead to a noteworthy therapy for colon cancer: bevacizumab (Avastin), which was approved by the FDA in 2004. But Avastin's effects on actual colon cancer in humans turned out to be more modest than what the manifesto science of anti-angiogenesis drugs would have predicted. If anything, Avastin is known more for its high price than for its clinical potency. And it certainly is no "cure for cancer."

Another takeaway from this episode is that Kolata's reporting was indeed accurate. She described the general reaction among Folkman's colleagues in terms of "cautious awe." It was Watson who decided to go out on a limb. The gap between his inaccurate forecast and Kolata's more measured reporting reveals the horizon bias in action. The broader lesson is that we will never be in a position to forecast cures, but we will have occasions of epistemic shift, when we can say, "*For all we know* now, cancer could be cured in 2 years."

When those occasions arise, we should put ourselves into the shoes of those in whom Folkman's discovery incited an eager sense of hope and epistemic possibility 23 years ago. The normal science needed to justify or deflate the relevant community's sense of awe had yet to be conducted. All anyone knew was that an experimental treatment specifically targeting tumor angiogenesis had never been done, and thus represented a step in an entirely new direction—a new horizon.

Because Folkman had identified a general truth about tumor growth, it seemed as though an entirely new path of research and clinical practice had been opened. From the epistemic perspective of 1998, knowledge of tumor growth would lead to the development of therapies that could halt or reverse it. We know now that it was not to be. But we also must recognize that for those in 1998, something approaching a cure seemed epistemically possible.

The Case for Pessimism About Civilizational Problems

Clinton's science advisers drew on good manifesto science and described a goal that was appropriate for the cancer scientists of 2000. The bet was that normal science would yield favorable results and encounter few obstacles; nobody was relying on a miracle to happen.

The problem arises when one uses otherwise good manifesto science as their basis for forecasting. Good manifesto science is functional, not prophetic. It tells us how to organize and pursue "stretch goals," not what we will encounter along the way. As the obligatorily sequential path to the final destination grows deeper, so does the uncertainty of any forecast. This is even more the case in biological systems, which tend to be far more complex and probabilistic than, say, the mechanics of space travel.

We know that normal science frequently produces unexpected experimental outcomes or observations. We thus should anticipate that the more steps there are in the agenda, the more likely the results are to deviate from our planned path. It is understandable that major scientific breakthroughs will be met with high hopes, especially among those who are intellectually and emotionally invested in the field. But we should expect that these hopes will be frustrated, not as a matter of principle, but out of experience.

Again, it is always *possible* that someone tomorrow will find a cure for cancer or a relatively easy fix for climate change. But there is no rational justification for expecting that. Since the age of Hippocrates, there have been many advances in our understanding and treatment of cancer. Many of these have fueled hopes of a cure, and many have indeed rid individuals of their cancers; but many of the traditional treatments are themselves potential causes of new cancers in the same individual.

There are many reasons why we should expect this pattern of disappointment to continue indefinitely. The first comes from oncologists and cancer researchers themselves. Though we have focused on optimists like de Grey and Watson, it is worth noting that many of those on the actual front lines of the War on Cancer are highly critical of the hype and overblown promises that have long attended cancer research. Many practicing oncologists have warned against believing the hype that so often emanates from their field. Their work makes clear that a great deal of cancer biology lies beyond our grasp, and that the institutions charged with waging war on the disease suffer from major structural flaws, bad incentives, and—at times— blinkered hubris.

"Arrogance. Overconfidence. Contempt." According to Columbia University oncologist Azra Raza, that is how Robert Weinberg described many of the cancer researchers that he encountered in the 1970s, and it is how she sees much of the field today. Raza urges her colleagues to accept that the problem remains extraordinarily complex and difficult despite the advances of the past half-century" (Raza 2019, p. 288).

Moreover, as a May 2020 editorial in *JCO Oncology Practice* warns, both the media and the pharmaceutical industry routinely contribute to the hype around new and expected interventions against cancer. Cancer care centers, too, have increasingly gotten in on the game, more than tripling their spending on advertising between 2005 and 2014 (Berry et al. 2020, p. 219). When the advocacy group Truth in Advertising looked at the top 50 cancer centers in terms of advertising spending in 2017, it found that 90% of had cherry-picked atypical positive patient results rather than explaining clearly what kind of results other patients could reasonably expect for themselves (TINA 2020).

Whenever we see a front-page story about some underappreciated visionary offering a revolutionary new treatment for cancer, we should remember these warnings from watchdogs and physicians on the front lines. Cancer is closely bound up with our own evolution as biological entities. As long as there are significant things to discover about ourselves and how we evolved, there will remain significant things to discover about cancer.

This is particularly true of the immune system, much of which we still do not understand. To defeat smallpox, we had only to figure out how to stimulate our immune systems to attack the variola virus. It will take much more to beat an enemy that emerges and spreads internally by dint of the fact that it has "learned" to evade our body's own countermeasures.

A final justification for lower expectations is psychological. Centuries of progress against cancer have not been for naught. We have discovered many effective therapies, and the fact is that there has never been a better time to receive a cancer diagnosis than right now. But the human condition is such that we today will always be dissatisfied with yesterday's impressive advances.

Since cancer cannot be eradicated like an infectious disease, the best we can hope for is steady, marginal improvement. Humans will always get cancer, and as long as those with cancer have different lived experiences than those without cancer, the disease will continue to represent a burden on society. Deciding precisely when we can declare cancer "cured" or "defeated" will be exceedingly difficult; and even if one generation arrives at a consensus on the matter, its progeny may beg to differ, demanding more.

In the meantime, experts and the lay public alike should bear in mind the distortionary effects of the horizon bias. We should all be skeptical of confident predictions issuing from the field, and even more so of sales pitches from those on the sidelines. We need manifesto science to guide broad, collective research efforts, and to keep hope alive. But we must be mindful of the effects it can have on even the best informed. Those with the deepest knowledge of cancer biology also have access to the widest scope of how-possibly explanations. They know both too much and too little to offer a useful forecast.

But what about the role played by technology? If computing power and the suite of available medical technologies are improving at an exponential pace, as we are often told, shouldn't *that* change the equation for the better? We will turn to that question later.

References

Berry, Leonard L., Timothy Keiningham, Lerzan Aksoy, and Katie A. Deming. 2020. When Cancer Centers Mislead Prospective Patients. *JCO Oncology Practice*, 1 May 16 (5): 219. https://doi.org/10.1200/JOP.19.00783.

Clinton, Bill. 2000. *June 2000 White House Event—Remarks by the President*. National Human Genome Research Institute., 26 June. https://www.genome.gov/10001356/june-2000-white-house-event. Accessed 26 Apr 2020.

De Grey, Aubrey, and Michael Rae. 2007. *Ending Aging: The Rejuvenation Breakthroughs that Could Reverse Human Aging in Our Lifetime*. New York: St. Martin's Press.

DeRose, Keith. 1991. Epistemic Possibilities. *The Philosophical Review* 100 (4): 581–605. https://doi.org/10.2307/2185175.

Deutsch, David. 2011. *The Beginning of Infinity: Explanations That Transform the World*. New York: Penguin Books.

DeVita, Vincent T., and Elizabeth DeVita-Raeburn. 2016. *The Death of Cancer: After Fifty Years on the Front Lines of Medicine, a Pioneering Oncologist Reveals Why the War on Cancer Is Winnable—and How We Can Get There*. Sarah Crichton Books.

Dray, William. 1957. *Laws and Explanation in History*. Oxford: Oxford University Press.

Gigerenzer, Gerd. 2019. *Gut Feelings: The Intelligence of the Unconscious*. New York: Penguin Random House.

Huberman, Gur, and Tomer Regev. 2001. Contagious Speculation and a Cure for Cancer: A Nonevent That Made Stock Prices Soar. *The Journal of Finance* 56 (1): 387–396. https://doi.org/10.1111/0022-1082.00330.

Kennedy, John F. 1961. *Address to Joint Session of Congress*, 25 May. John F. Kennedy Presidential Library and Museum. https://www.jfklibrary.org/node/16986. Accessed 11 Nov 2022.

Kolata, Gina, 1998 Hope in the Lab: A Special Report; A Cautious Awe Greets Drugs That Eradicate Tumors in Mice. *The New York Times*, 3 May.

Kuhn, Thomas. 1962. *The Structure of Scientific Revolutions*. Chicago: University of Chicago Press.

Laplace, Pierre-Simon Laplace (translated by Frederick Wilson Truscott and Frederick Lincoln Emory). 1814/2015. *A Philosophical Essay on Probabilities*. Chronicon Books.

Niiniluoto, Ilkka. 2019. Scientific Progress. *The Stanford Encyclopedia of Philosophy*, 16 October. https://plato.stanford.edu/archives/win2019/entries/scientific-progress. Accessed 10 Mar 2020.

Raza, Azra. 2019. *The First Cell: And the Human Costs of Pursuing Cancer to the Last*. New York: Basic Books.

Resnick, David B. 1991. How-Possibly Explanations in Biology. *Acta Biotheoretica* 39: 141–149. https://doi.org/10.1007/BF00046596.

Sinclair, David. 2021. *Smartless* (podcast) 21 February.

TINA. 2020. *Summary of Action. Truth in Advertising*. https://www.truthinadvertising.org/cancer-centers-summary-action. Accessed 25 July, 2020.

Wade, Nicholas. 1997. Tests on Mice Block Defense by Cancer. *The New York Times*. 27 November.

Wong, C.H., K.W. Siah, and A.W. Lo. 2019. Estimation of clinical trial success rates and related parameters. *Biostatistics* 20 (2): 273–286.

Chapter 4
The End of Disease!

Contents

Since the War on Cancer is now 50 years old, it offers a data-rich counterexample to consider alongside confident claims about the imminence of technological solutions to civilizational problems. Simply put, Susan Sontag's description of the problem in 1978 still rings true today: "in an era in which medicine's central premise is that all diseases can be cured," cancer remains "intractable and capricious" (Sontag 1978, p. 677).

Like the stereotypical pageant queen who wants "world peace," politicians and techno-optimists are perpetually tempted to promise a cure for cancer. Hence, on June 11, 2019, Joe Biden, then a presidential contender in the Democratic Party primary, said that, "If I'm elected president, we're going to cure cancer." To be sure, this was not just empty campaign rhetoric, for Biden had lost his son Beau to a brain tumor in 2015. Still, he was on well-trodden ground, and a few days later, Donald Trump, in typical fashion, promised even more: "We will come up with the cures to many, many problems, to many, many diseases … including cancer." (McDonald 2019).

It is no secret why politicians would trot out this cliché, whether out of a genuine commitment to winning the War on Cancer or out of political expediency. Since 1999, cancer has been the leading cause of death for all Americans under the age of 85 (Plutynski 2018, p. 216; Harding et al. 2018). Around 40% of American men and 39% of American women will confront the disease in their lifetimes. And despite all of the technological means at our disposal, 20.5% of American men and 17.9% of American women will die from the disease (ACS 2022).

N. Agar et al., *How to Think about Progress*, Library of Ethics and Applied Philosophy 42, https://doi.org/10.1007/978-3-031-68938-3_4

Though there have been improvements in preventing, diagnosing, and treating certain forms of cancer, progress overall has fallen far short of what has been promised. In 1971, US President Richard Nixon announced an "intensive campaign to find a cure for cancer," declaring that, "The time has come in America when the same kind of concentrated effort that split the atom and took man to the moon should be turned toward conquering this dread disease" (NCI 2022). By the end of that year, he had signed the US National Cancer Act, officially declaring war on the disease.

Victory, everyone assumed, would come from finding *the* cure for cancer. Nixon's speechwriters even offered what they saw as a viable deadline for such an achievement: 1976. Success within just 5 years certainly would have made up for the Watergate Hotel break-ins and much more. And yet, finding *the* cure for cancer is not the same thing as curing some cases of cancer in some patients—something we've long been able to do. People were still dying from cancer in 1976, and cancer remains stubbornly uncured and undefeated 50 years later.

Two Faces of Progress

Will the promises issued by Nixon's twenty-first-century successors fare any better than his did? Looking back from the 2020s, the idea that cancer might have been cured by 1976 seems absurd. It doesn't feel like we are much closer to that goal now than we were then.

One reason for confusion about progress against cancer is an ambiguity in what it means to make progress against the disease. The Finnish philosopher of science Ilkka Niiniluoto points to two different but complementary perspectives on progress. There is progress *from* some initial state, and there is progress *toward* some future condition. If you're driving to a nearby town, you can measure your progress from your point of origin ("How far have we gone?"), or you can measure it in terms of distance that remains ("How much farther?") (Niiniluoto 2019).

The War on Cancer has certainly seen much progress from Nixon's declaration. If you had to choose a time to be diagnosed with cancer, you would pick today over 1971. You would also expect that 2041 will be a better time to be diagnosed than today. If we focus on this first sense of progress—the distance traveled *from*—then the years since 1971 could be celebrated. There are so many effective treatments that would not have existed but for Nixon's declaration of war and the funding it unleashed.

It is thus the complementary sense of progress—the distance left to cover—that induces disappointment. We are collectively aiming for a destination represented either by the cure or some other outcome that we could collectively accept as constituting victory in the War on Cancer. As we will see, our assessments of progress *toward* the cure for cancer are complicated by the fact that we can only guess at how difficult the remaining task will be.

A future society that possesses the cure for cancer will presumably know how difficult it was to cure. But we lack that knowledge. It's as if we've set off on a trip to a town without any idea of how distant it is. We can make accurate assessments of progress from our point of origin, but our assessments of progress *toward* are grounded in the hope that our destination won't be too far away. The distance left to travel could be 5 miles, or it could be five trillion miles. We do not currently know.

A failure to distinguish these two senses of progress explains why there are dueling assessments about how things are going in the War on Cancer. On slow news days, journalists who are interested in progress *from* tend to write "We're-winning-the-war!" stories focusing on exciting new therapies such as mRNA cancer vaccines and other immunotherapies. Meanwhile, those focusing on progress *toward* tend to write "We're-losing-the-war!" stories, pointing out that there is no end in sight even after a half-century of constant effort and billions of dollars spent.

Still, we do have at least some sense of what an imminent end to the War on Cancer should feel like. In his overly optimistic 2003 assessment of the war in Iraq, Donald Rumsfeld referred to the enemy bitter-enders "who would fight to the end." When we have gotten to the point of fighting an opponent's bitter-enders, we can presume that even they know their cause is lost. That was doubtless the case for bitter-ender German defenders of Hitler's bunker who fought determinedly even as Soviet forces closed in. But in the War on Cancer, by contrast, it does not feel like we have reached this stage. We are not down to eliminating a few occasional bitter-ender tumors.

What we do know is that cancer is a deeply complex emergent phenomenon based on a variety of factors including genetic mutations, the "microenvironment" (the surrounding tissue in the body), environmental carcinogens, viral infections, and other variables. These inherent complexities, warns the late cancer researcher Kenneth L. Mossman, "limit what we can ultimately know and understand about the human body and disease." Cancer medicine, in particular, faces what Mossman calls a "complexity paradox – the more we learn about complex systems the more questions we ask." Owing to the inherent uncertainties, he suspects that, "we can never know everything there is to know about cancer and other chronic degenerative diseases" (Mossman 2014, p. xi-xii).

Many of the biggest advances against cancer do not come from discoveries that reach into the disease's inner workings. Indeed, our biggest victories seem to come from prevention (particularly the decline in tobacco smoking) and early diagnosis, rather than from technological breakthroughs in treatment. The expenditure of tens of billions of dollars on the war has not yielded progress comparable to that made against stroke and cardiovascular disease, where the age-adjusted death rates have plummeted.

Is this a choice between viewing progress against cancer as a glass half-empty or half-full? Optimists can go with the half-full perspective and point to all of the exciting therapies that wouldn't have existed had Nixon not gone to war. Pessimists can focus on the half-empty side and point out that there is no end in sight. But anyone making this implicit choice should understand that the glass is both half-full and half-empty.

Such a dispute is not easily resolved. Part of the problem lies in the way we think about cancer research. Too many cancer scientists are overly focused on the publicity that comes from promising "the cure." They understand that this language works when presenting grant applications, some of whose assessors won't be experts in cancer research. The experts on the panel may be excited about research that could produce a compound that might up-regulate expression of the PTEN gene—mutations of which often show up in lung cancer, prostate cancer, and melanoma. But the aspects of the proposal that will make it exciting to the other interdisciplinary members of the panel will be claims that research on this jumble of letters could—*just possibly*—lead to a cure for cancer. That, anyway, is how successful applications for funding tend to be marketed.

When the cancer researcher Vincent DeVita ended his long career studying combination chemotherapies and sat down to write a book, he chose a determinedly progress-*toward* title: *The Death of Cancer* (DeVita and DeVita-Raeburn 2016). Though he had many *past* advances against cancer to celebrate, his publisher probably wanted a title oriented toward the future. The result was a title—and overall attitude—that overlooks the horizon bias, setting up readers for inevitable disappointment.

As Siddhartha Mukherjee shows in *The Emperor of All Maladies*, the struggle against cancer has been a story of repeatedly dashed hopes, exhibiting a now-familiar pattern of overwrought excitement toward some new discovery, followed by the inevitable let down (Mukherjee 2011).

Remember, everyone got a dopamine hit from Clinton's line about "our children's children" only knowing the term "cancer as a constellation of stars." That was predictably followed by disappointment. More than two decades later, some people's unlucky grandchildren are still dying from cancer, and we are now getting our dopamine hits from the news that AI-assisted protein folding "could dramatically change how we fight disease" (Griffin 2020).

If we look back to 1963, following the discovery that chemotherapy could be effective in treating certain forms of cancer, we can hear the director of the National Cancer Institute, Kenneth Endicott, predicting that, "The next step – the complete cure – is almost sure to follow" (NIH 2020). That sounds a lot like James Watson 35 years later, and like David Sinclair almost 60 years later.

What Is the Cure for Cancer?

Hearing such confident predictions, the question one must ask is what it would even mean to come up with *the* cure for cancer or to win the war on the disease. The answer is not as obvious as it first seems, because we have in fact slipped into the realm of metaphor.

While Nixon and his generation declared war on cancer, Trump and Biden promised to cure it. At first blush, the two images could scarcely be more different. In war, one orchestrates violence against a perceived enemy, generally through the deployment of armed forces, the dropping of bombs, and so forth. The goal is to eliminate the enemy entirely, or to extract a surrender. To "cure," by contrast, is to care for, "to restore to health," following from the Latin *curare* (which long had religious connotations: the job of the medieval curate or curé was to care for souls). We will return to these metaphorical responses to cancer later in the book.

An individual with cancer can hope for a literal cure, because she can see that many other people with certain types of cancer have rid their bodies of the disease and gone on to live full lives. Squamous cell carcinomas, for example, are usually curable. When a squamous cell carcinoma is excised from the top of your head, you can count yourself cured of that cancer. In your specific case, the scalpel will have been *your* cure. Ever since the first forager used a serrated stone fragment to saw off some unsightly skin cancer, individuals have been cured in this way. But if this is what we meant by finding "the cure," we would have already declared victory many times over.

Applying to our collective challenge a term that describes individual experiences is what philosophers would call a category error. Consider the annual flu. Humans as a species are susceptible to influenza (a viral infection), and many of us occasionally get it. But humankind does not get the flu. To say that we might cure humankind of cancer in the same way that we might cure individuals of cancer is to describe a result that is conceivable but actually impossible.

With a disease like smallpox, we can equate "eradication"—defined as zero new cases—or "elimination" of the disease-causing agent (variola) with the cure. The last diagnosis of a natural case of smallpox occurred in 1977, and the World Health Organization was able to declare the disease eradicated in 1980 (WHO 2022). After 40 years, we can be confident that the variola virus is not lurking in the bodies of anyone who is alive today, even among those who survived smallpox in the 1970s.

Even in this case, the cure is contingent. There will never be another smallpox case as long the last remaining variola samples remain under tight security in the United States and Russia, and as long as humans do not come into contact with some hitherto unnoticed sample, and as long as no malevolent genetic engineer recreates the virus in a lab. If one were to contract a new case, we would suddenly remember that there actually isn't a "cure" for smallpox. Nonetheless, we can safely assume that our grandchildren will know the disease only as the name of some past scourge that one finds in history books.

In any event, the eradication-as-cure option is not available for cancer. Though some cervical and other cancers can be triggered by exposure to viruses, eradicating those viral infections will not eradicate cancer. Humanity could eradicate smallpox because the variola virus is an external enemy. Vanquishing that disease took decisive, collective action, but it did not require any fundamental revisions to our biology.

Sidney Farber, the famous twentieth-century pediatrician and pioneer of chemotherapy, first raised the idea of a "universal cure" for cancer in 1962 (Mukherjee 2011, p. 155). Farber changed the way we think about cancer when he discovered that a drug could shut down the production of folic acid in leukemia patients, resulting in temporary remissions among a population that had hitherto been considered untreatable.

Over time, chemotherapy would evolve to include an ever-growing alphabet soup of drugs designed to kill off every last cancer cell (along with all other cells that undergo rapid replication, from hair to bone marrow). The thinking went that if certain cells were resistant to Drug A, they would succumb to Drug B, and if not that, to Drug C.

The problem with this approach to a "universal cure" is that it is at odds with our very essence as evolved beings—that is, as the products of evolution and of constant cellular "becoming." There are plenty of times when a cocktail of drugs will kill off a patient's tumor. But if a single cancer cell proves resistant, it is that cell that will go on to replicate *ad infinitum.*

If we were machines assembled out of a finite collection of components, we could presumably confirm that each part is factory-perfect before it is placed into us. We could promptly replace faulty components and download system updates to ward off new carcinogenic threats that will emerge.

To eradicate cancer, we would need to alter our very nature. As mammals, we humans grow from a single cell, and are host to a constant, intricate choreography of trillions of cell divisions. Nothing is ever static; what holds today may have changed by tomorrow. While we owe our very existence to our cell divisions, we are also at the mercy of the replication errors that naturally and inevitably occur. Most of these are duly corrected by cell-repair mechanisms in our own bodies. But some mutations allow a cell to evade these defenses.

Hence, the idea of a universal cure for cancer belongs in the same aspirational realm as a universal cure for death. We can imagine it in the context of certain static conditions, but as soon as we try to grab ahold of it in the real world, we realize that it is constantly moving beyond our reach.

As such, the computer scientist Daniel Hillis believes that we should change the grammar of how we talk about our experience with cancer. Rather than focusing on "cancer" as a noun, Hillis thinks we would do better to use the verb form: "cancering." Cancer is not something that "you have," Hillis argues. "It is something you do. Your body is probably cancering all the time. What keeps it under control is a conversation that is happening between your cells, and the language of that conversation is proteins" (Hillis 2010).

In other words, we're all cancering right now, albeit to different degrees. If you have colon polyps, that is an indicator of a certain level of cancering, even if your doctor is considerate enough to have spared you the "C" word. If you smoke or spend a lot of time in a tanning bed, you have chosen to engage in more cancering than you otherwise might have done. When one's cancering gets out of control, one dies from it. But those who died from smallpox were never "smallpoxing."

There's the rub. As a noun, smallpox is an appropriate target for eradication; as a verb, cancering is not. The fact that we are living, aging beings means that, as Anya Plutynski writes, "there is always the potential for cancer to come about" (Plutynski 2018, p. 217). The proto-oncogenes that can mutate into cancer-causing oncogenes are *normal* genes that play necessary functions in the body. We could try to excise all of our proto-oncogenes with powerful new gene-editing technologies, but then we'd be dead. It would be like trying to prevent heart attacks by removing one's arteries.

The Horizon Bias vs. Sci-Fi Cures

As the Emperor of All Maladies, cancer looms large in humanity's collective nightmares. The cure thus frequents the lists of possible technological advances that futurists can trot out without much thought. There, it is joined by things like "the end to aging" and colonizing Mars. But the horizon bias will continue to fuel disappointment about our inadequate progress toward curing cancer. It leads us to be far too impressed by the fantastical treatments that are promised by those pursuing a manifesto science of cancer.

If we were to go back in time to share today's cancer statistics and therapies with Nixon's scientists, what would their assessment be? For starters, we can assume that they would be amazed by ingenious new immunotherapies and other biotech interventions that tap directly into the molecular bases of our immune systems.

Even to the most sophisticated practitioners of the early 1970s, these marvels would have looked like something out of science fiction. One can imagine a 1971 sci-fi novel set in 2021, depicting a world where futuristic immunotherapies could direct the characters' immune systems to blast away any incipient cancer. To 1971 eyes, such treatments would seem to augur the cure. They would have made perfect sense as manifesto science for 1970s-era readers who wanted to become cancer researchers.

But then we would have to show the people of 1971 the data about cancer incidence and death rates. They would realize that none of the powerful new therapies constitute anything close to the cure. They might then realize that half a century of normal science had turned up many unforeseen obstacles and complications.

Today's futurist cancer fixes generally observe the laws of physics and logic, which makes them feel as though they are imminently achievable with enough resources. All follow from the theoretical physicist David Deutsch's dictum that human beings are universal constructors capable of "transforming anything into anything that the laws of nature allow." The implication is that "matter, energy and evidence are the only requirements that an environment needs to have in order to be a venue for open-ended knowledge creation" (Deutsch 2011, p. 59, 62).

In the case of cancer, we know that it is a problem of unregulated cell growth, and that one solution is to kill those cells. We also know how to kill cells—humanity learned that when it discovered fire. The challenge is to kill the *right* cells. Our current methods—surgery, chemotherapy, radiation—have proved insufficient and unreliable. But, if we take the techno-optimist's view, it is easy to imagine new methods that could achieve the same end goal more effectively, indeed perfectly.

At the time of this writing, there is a great deal of excitement about immunotherapy—or immuno-oncology—an area of research that was recently considered a dead end. Unlike long-established treatments such as chemotherapy—which simply destroys rapidly growing cells, cancerous or otherwise—immunotherapies seek to disrupt cancer cells' ability to evade destruction by the immune system.

One vocal proponent of this line of research is Ray Kurzweil, a futurist and former child prodigy inventor who predicted, in 2016, that "By the 2020s we'll start using nanobots to complete the job of the immune system." And as these technologies "gain traction in the 2030s," he adds, "nanobots in the bloodstream will destroy pathogens, remove debris, rid our bodies of clots, clogs and tumors, correct DNA errors and actually reverse the aging process. One researcher has already cured type 1 diabetes in rats with a blood-cell-size device" (Hochman 2016). Kurzweil's techno-optimist perspective treats cancer as a straightforward fix. The underlying assumption is a hallmark of modernity: if we understand the nature of the problem, we can fix it.

The proposal seems eminently plausible, judging by all the world-changing innovations that have come out of Silicon Valley (and Shenzhen) in recent years. After all, no part of the plan violates the laws of physics or logic.

There are similar imagined techno-fixes for climate change. Presumably a different design of nanorobot to the one Kurzweil imagines retargeting our immune systems to take out any tumor can be dreamed up to be sprayed into our upper atmosphere to take out excess carbon. Once the levels of carbon are restored to pre-Industrial Revolution levels, the nanorobots could presumably be deactivated by a radio transmission to prevent them from completely stripping the planet of its atmospheric carbon.

The problem with techno-optimistic solutions to the big Cs is that the confidence they inspire—and the dopamine hit they deliver—make us worry even less about the burning of coal and other cheap and efficient forms of energy in the meantime. If human ingenuity knows no bounds, why should we give up cheap electricity (or potentially carcinogenic chemicals, for that matter). Cheap energy is good for economic growth and therefore material well-being, right? If anything, we should be trying to maximize the returns of coal. Even if it exacerbates climate change in the short term, the long term will bring powerful new mitigation technologies the likes of which we cannot even imagine today.

Given the stakes, we should interrogate the assumptions that underpin this perspective. Insofar as cancering is bound up with our biological existence, eradicating cancer requires that we transform the human condition. Not surprisingly, there are a variety of futurist proposals to defeat cancer that focus on ways to change our biological fundamentals.

Like the technofixes for climate change, these ideas have the added advantage of not requiring us to make any sacrifices. With a reliable technological solution to cancer, one could smoke as many cigarettes as one wants. The problem is when progress toward technological solutions arrives at a slower pace than any lung cancers triggered by those celebratory smokes.

The laws of physics offer a wide range for imaginative conception, from time travel to the creation of alternate dimensions. When it comes to overcoming biological problems like cancer—and the perishability of the human body more generally—futurist philosophers have taken a keen interest in mind-uploading, in which we would transfer all of the "data" from our brains and bodies into suitably powerful computational devices. At this point, we might forsake our cancering biological selves entirely, or we might stick with them while keeping a digital reincarnation of ourselves in the cloud.

Make no mistake, advocates of mind-uploading present it as a serious possibility, following from contemporary neuroscience and a much longer tradition in the philosophy of mind. Once one comes to view the mind as a set of neuro-processes and algorithms that are distinct from the body, it stands to reason that the identity-relevant data therein could be translated into the language of computer code. Ironically, if we were to transform into robots, we may suffer from entirely new diseases or bugs; but at least we wouldn't get cancer.

Another option for re-engineering our fundamental nature focuses less on the mind than on the body; but it is no less ambitious. A leading exponent of this approach is Aubrey de Grey, whom we met in the last chapter.

We should recognize mind-uploading and de Grey's WILT program for what they are: science-fiction cures for cancer. They belong in the same moral universe as technologies that transport vacationers to the dwarf planet Pluto. If one devotes enough time and energy to describing how that might work, extrapolating from current technologies and reasonable expectations of technologies on the horizon, it is not all that difficult to imagine. But pitch that highly detailed, scientifically coherent plan to serious, mature investors, and you will likely be met with a blank stare. Your perfectly plausible prediction will have zero effect on today's tourism industry.

To be sure, as fully articulated examples of hard sci-fi, WILT and mind-uploading are in a different category than Elvish herbs or magic spells. Both lines of thought would be right at home in the work of the Chinese master of hard sci-fi Liu Cixin, author of *The Three Body Problem*. But neither deserves serious consideration by those trying to solve humanity's cancer problem.

As long as cancer remains the scourge that it is, those selling utopian cures will have an eager audience. We may not be able to ignore them, so we should opt for an attitude of deep skepticism and caution. If your gambling friend wants you to stake her for the roulette table, she will probably try to entice you with a vision of big winnings, promising you a cut. She'll make it sound as if she cannot possibly leave the table with less than she had coming to it. But before you stake your friend, you should stop and seriously consider the odds. Unless she has found a way to fix the roulette wheels, your investment is likely to go the way of the majority of Las Vegas investments: into the casino's coffers.

De Grey is also inviting his audience to stake him on a wager (Friend 2017) The outcome he has promised is neither logically nor physically impossible, from what we can tell. But that doesn't make his project a good bet. Rather, de Grey seems to be betting on his audience's "optimism bias," whereby people tend to "overestimate the likelihood of positive events and underestimate the likelihood of negative events" (Sharot 2011a, p. R941). When people understand that hundreds of billions of dollars have been thrown at a problem like cancer, they naturally assume that something will come from it. Likewise, with cancer itself, most people naturally assume that, though it is a leading cause of death, it's the kind of thing that happens to someone else, not to them.

According to the Israeli neuroscientist Tali Sharot, the optimism bias is, "one of the most consistent, prevalent, and robust biases documented in psychology and behavioral economics," and likely evolved because it served our interests (Sharot 2011a, p. R941; Sharot 2011b). In the nasty, brutish competition for scarce resources or mates, those who were optimistic about their chances likely competed harder than those who were not. (As the saying goes, you can't win if you don't try.)

In a competition among individuals, optimism on the part of every participant is most likely harmless, or even beneficial if it improves performance. But an optimism bias on the part of society as a collective can be dangerous, leading us to ignore threats like climate change because we assume that someone somewhere will eventually come up with a solution before it is too late. And perhaps they will. The point is not that we won't find a fix. It is that we are more likely to generate a solution if we do not all assume that someone else will do it for us.

We should interrogate the ethics of an approach that puts us in the position of needing a "Hail Mary" solution. We should recognize that the evangelists of sci-fi fixes are pandering to our optimism bias (or to desperation, in the case of cancer patients). As is always the case with confident predictions, we should ask who is making them and what those people stand to gain. Politicians want our votes, and entrepreneurs want our money.

A Cure for Diabetes, Too?

Cancer is not alone among diseases we expect to see cured. One of us—Agar—was diagnosed with type 1 diabetes in 1990. No Emperor of All Maladies, diabetes belongs to a lower rank of the disease nobility. A cure for it thus lacks the civilizational significance of a cure for cancer. Nonetheless, Agar has eagerly anticipated a cure for more than 30 years. After the shock diagnosis, he was assured that diabetes was essentially a solved problem; that he would have to deal with insulin injections and finger pricks for 5 years, at most. Clearly, his doctor at the time was captured by the horizon bias.

Frederick Banting and Charles Best isolated insulin in the early 1920s, whereupon an invariably fatal disease for many diabetics suddenly became treatable. In Michael Bliss's authoritative 1982 history, *The Discovery of Insulin*, we learn that:

"The discovery of insulin … was one of the most dramatic events in the history of the treatment of disease. Insulin's impact was so sensational because of the incredible effect it had on diabetic patients. Those who watched the first starved, sometimes comatose, diabetics receive insulin and return to life saw one of the genuine miracles of modern science. They were present at the closest approach to the resurrection of the body that our secular society can achieve, and the discovery of what has become the elixir of life for millions of human beings around the world" (Bliss 1982, p. 11).

After Banting and Best's breakthrough, all that was needed for a cure was to automate insulin injections. It seemed simple, in principle, because of the horizon bias.

For their part, the scientists who first isolated insulin downplayed its significance, because they anticipated much more progress to come. In his 1923 Nobel Prize lecture, Banting explained that, "Insulin is not a cure for diabetes; it is a treatment. It enables the diabetic to burn sufficient carbohydrates, so that proteins and fats may be added to the diet in sufficient quantities to provide energy for the economic burdens of life" (Banting 1925). Banting's goal was to turn his discovery into a cure, or at least into something that could closely approximate one. Indeed, a June 9, 1921, entry in his notebook outlined a plan to graft pancreatic tissue from dogs into diabetic patients (Shapiro 2002, p. 1398).

Banting never got around to pursuing that ambition. But it is obvious that he saw a clear path forward. Insofar as his plan seemed eminently feasible, it inspired just as much confidence in the diabetes researchers of his generation as had advances in islet-cell transplants for researchers in the 1990s.

It was this eager hope for a straightforward cure that inspired the creation of the Flame of Hope, an eternal flame completed in 1989 in London, Ontario, Canada. The Flame is to be extinguished only when a cure is found. The relationship between work to improve the treatment of diabetes and work to find a cure is less straightforward than it might have seemed to the builders of the Flame of Hope.

The Flame was not intended to be a tragic reminder of our failure to make it across the finish line. The idea was that it would be dramatically extinguished within a few years. But the path from marginally improved treatments to a cure turned out to be less straightforward than expected.

The Flame of Hope was actually extinguished in 2020, but not because diabetes had been cured (Lancione 2020). The vandals that put out the Flame of Hope may have misunderstood the direction of causation between extinguishing the Flame and curing diabetes.

Visions of imminent breakthroughs against diabetes are based on the same kind of manifesto science that has long underpinned promises of a cure for cancer. Diabetes has not been cured, and by 2019, the biggest story concerning the disease was that a growing number of diabetics can no longer afford insulin, owing to price gouging by the pharmaceutical industry (Kreizman 2019).

To be sure, there have been significant advances in the treatment of diabetes since Banting and Best's breakthrough. A focus on progress *from* might point to the recent advances of insulin pumps and continuous glucose monitoring devices that replicate some of the functions of a healthy pancreas. To extinguish the Flame, one would have to devise and make universally available a treatment that renders Type 1 diabetics' experiences indistinguishable from that of non-diabetics.

For someone dying a slow diabetic death in the last days before Banting's discovery, insulin therapy must have seemed almost as good as a cure. But that's not how diabetics view it today. Below, we will explore the psychological bases of our predicable disappointment in the quest for cures.

Hedonic Normalization

Short of turning ourselves into robots, we cannot eradicate cancer, because it is woven into our evolutionary essences. Nonetheless, a feature of human psychology continues to give life to the hope that cancer will finally be cured. As a result of what we call hedonic normalization, progress against cancer tends to move back the goalposts of what we would collectively accept as "the cure." What seems good enough for us in 2024 will predictably be inadequate for people in 2044.

Hedonic normalization thus can be understood as "the propensity for human beings to form goals that are appropriate to the environments they experience as they come to maturity" (Agar 2015, p. 3). As we've seen, the horizon bias especially afflicts those with the greatest knowledge about the relevant science. By contrast, hedonic normalization is a limitation on progress that applies most to those who receive the latest therapies for cancer or diabetes. These imperfect treatments ineluctably lead them to hope for something even better.

Hedonic normalization perpetually limits our capacity to appreciate the benefits of new technologies over time. Because it operates at the generational rather than the individual level, it is a constant psychological companion to technological progress. Hedonic normalization suggests that what one generation might accept as a notable achievement, later generations might find disappointing. Once a technology has been widely embedded in the culture, the first adopters grow used to it, and the next generation takes it for granted completely, just as we in advanced economies already do with electricity, air travel, and flush toilets.

There is an important distinction to be made between the inter-generational phenomenon of hedonic normalization and that of hedonic adaptation, which describes how we as individuals respond to positive and negative events in our lives. The classic case study involves a lottery winner who immediately undergoes a period of extreme happiness but experiences declining hedonic benefits from her windfall over time (Brickman et al. 1978). A similar process is said to follow from negative events such as a permanently disabling injury.

Happiness research in general is a contested field, owing to its reliance on subjective, often self-reported data. Nonetheless, with ample aggregate data, there are enough correlations to have attracted more economists, psychologists, and other social scientists into the field.

Some researchers believe that hedonic adaptation involves a complete return to a happiness "set point," implying that with enough time, unexpected benefits or costs will leave one no better or worse off than he was before. The idea is that someone who becomes paralyzed in an automobile accident will initially suffer extreme misery, but will eventually grow accustomed to his new way of experiencing the world (Diener et al. 2006; Brickman et al. 1978).

Others agree that such adaptation occurs, but doubt that it runs in a complete cycle back to a pre-set baseline. The accident victim will not remain at the height of misery indefinitely, but nor will he ascend back up to the same level of subjective well-being. Rather, he will be reminded of the capabilities he has lost every day when he gets out of bed.

In any case, most people know from their own lives that at least some degree of hedonic adaptation occurs. Such recognition is deeply embedded in the broader culture. Consider the clichés that follow misfortune: "This too shall pass"; "Don't worry, time heals all wounds." Everyone grows up learning to be grateful for what they have. Americans have a major holiday dedicated to the ritual of giving thanks for all of the things they have taken for granted throughout the year.

There is a paradox to how we approach technological progress. As individuals who can imagine the possibilities opened up by driverless cars or age-reversing "rejuvenation" drugs, we should ask what these technologies will do for our own well-being. The mania that greets each release of a new iPhone seems incompatible with the modest objective improvements that come with each new model.

True, the first affordable, safe driverless car would surely deliver a more significant improvement for human well-being than an iPhone 16 does, especially if one already owns an iPhone 15. And yet, we can already foresee that the first driverless car ultimately will add only modestly to individuals' subjective well-being in the long run, because the people who experience the initial excitement about a major new advance will be replaced by people are hedonically normalized to it.

Following a predictable pattern, the first passengers will be mightily impressed. But as time passes, the cabins of driverless cars will increasingly be filled with people who take their setting for granted as much as post-millennials do on the internet. These future people naturally would abhor the idea of relinquishing their technologies' advances, but they also would feel rather complacent about them.

As individuals, we can pine for the drug-like effect of technological progress, and we can even satisfy it temporarily from time to time. But the implication of hedonic normalization is that our collective thirst will never be quenched.

One can see this pattern clearly in cancer research. Sydney Farber achieved miraculous remissions of leukemia with chemotherapies that continue to be therapeutically effective. And yet, today's cancer patients tend to submit to chemotherapy not with a sense of wonder but rather with resignation. Chemotherapies have become ever-more clinically effective since Farber's day, but these improvements have had a diminishing effect on our expectations of progress against cancer. Farber's objectively inferior treatments packed a much greater hedonic punch.

Such is the nature of progress. With each passing generation, the expectations placed on technology and human ingenuity are renewed. Aspirations achieved today will be regarded as wholly insufficient by members of the society of 2100. Few in the rich world could imagine going about their days without air conditioning, yet that is what the bulk of humanity still does, and always has done. Because so many of our material and technological advances have been inherited, we take them for granted and demand more. If there ever comes a time when people do live to a thousand years, a millennial lifespan will no longer be considered adequate.

To be sure, we can certainly appreciate today's technologies intellectually, when we consider them in a historical context. But, hedonically, we are fundamentally bound up with our own conditions. Not only do "digital natives" fail to experience the same excitement as "digital immigrants" when they use digital technologies, but they also are more dependent on them. A digital native who is forced to surrender access to the internet would probably suffer a much greater blow to her well-being compared to a digital immigrant. An older adult would still remember the days before the internet, when people had real friends instead of Facebook friends. For them to feel the same sense of deprivation as the digital native feels, they would have to go further back in time, giving up color television or the polio vaccine.

Because digital natives are normalized to the digital age, they have no real sense of what life was like before. Even though they can "google it," doing so will not afford them a genuine understanding. To come closer to a sense of truth in this context requires imagination. Throughout history, humans have repeatedly invented technologies that (we can safely assume) were thrilling for those who first experienced them. But the sense of novelty doubtless started wearing off during their own lifetimes, and it wasn't available at all to their grandchildren.

This indelible feature of history has important implications for any discussion of progress. In order to say that our subjective well-being benefits significantly from the technologies that we were born with, we must adopt a preposterous assumption about the happiness of people in the past.

Suppose that you were suddenly teleported to a Roman city in the first century CE. You have no modern technologies with you, and no return ticket. After the initial wave of shock or excitement, the lack of familiar modern contrivances would almost certainly start to have a depressive effect on your well-being. Before Mount Vesuvius erupted and buried Pompeii under a sea of molten lava and ash, quotidian life in the city would probably have been rather miserable to a time traveler from the future. According to the classicist Mary Beard:

> Stepping down onto the road surface risked more than a twisted ankle; it most likely involved treading into a smelly mixture of animal dung (each horse producing up to 10 kilos a day), rotting vegetables and human excrement – which was, just to complete the picture, no doubt covered in flies. (Beard 2010, p. 56)

The implication is not that ancient Pompeians actually enjoyed these features of city life. We know that they did not, hence the need for public notices warning: "Shitter – make sure you keep it in till you've passed this spot." The difference is that the rotting vegetables, animal dung, human excrement, and flies would have much less of a deleterious effect on the Pompeiian's sense of well-being than on the time traveler's. To city dwellers up until the age of the automobile, shit-filled streets were akin to standstill traffic or online spam today.

The same basic caveats apply to future technologies. Supposing that human civilization does not degenerate into the kind of post-apocalyptic scenario depicted in *Mad Max*, a time traveler visiting the twenty-third century from the present would doubtless be awed by what she encounters, be it advances in space travel or the eradication of diseases and methods of extending human life. But the people she meets will be hedonically normalized to these technological wonders.

Or, consider another scene that can be summoned up on any computer screen right now: the interior of the starship *Enterprise* in *Star Trek*. How exciting would it be to wake up and find oneself surrounded by technologies that allow for human teleportation, warp-speed travel, and the like? A diabetic could go straight to Chief Medical Officer Dr. Leonard "Bones" McCoy and receive a seemingly miraculous cure to his condition. But nothing like the ecstatic joy he feels would be mirrored by the other crew members. They would look at his disease as a historical curiosity, much as we do with the plague.

What do these journeys of attitudinal time travel tell us? Based on our own first-hand experience of the present, we can assume that people in the past were not radically more or less "happy," on average, than we are today; and we can presume the same about people far in the future. The upshot is that technological progress can never boost our subjective well-being as much as its evangelists would like us to think it will. Even if technological progress didn't create new problems of its own, even if it didn't make human civilization ever more dependent on an increasingly complex and fragile technological architecture, it still could never satisfy the demands that we make of it.

The common mistakes that we make when thinking about the future could be dismissed as harmless quirks of the human condition, except that they tend to distort our priorities. The more that we focus on technological fixes to our problems, the less attention we pay to all other potential solutions.

What are the implications of hedonic normalization for our quest to cure cancer? We have seen that the easily imagined goal of eradication is incompatible with our biological natures. We can eradicate an external foe like smallpox but not an internal adversary like cancer. Short of turning ourselves into robots, we cannot achieve eradication.

Insofar as it results from the reproduction of cells, cancer is an endemic feature of life. Were we to find a way to extend cancer survival by 10 years, on average, we would have good reason to celebrate the achievement. But if we checked back in 20 years, we may well find the generation of 2044 wondering why humanity still has not won the War on Cancer. As we will see, it will take more than new therapies to win the war.

If we've eradicated smallpox then there is nothing to hedonically normalize to. Smallpox would, for all but historians, be justifiably out of mind. Even as we make progress, there will always be cancer. Over time, hedonic normalization will continue to push back the goal posts in the War on Cancer or the search for the cure. We will always have cancer to worry about. It will continue to be a terrible way to die. We will continue to bemoan that fact that the war is far from being won.

We would submit that the first step to curing cancer is not to look for a *deus ex machina* from Silicon Valley. It is to recognize that "the cure" exists only as a meta-phor. The cure for cancer is not a fixed, achievable goal, but rather a way thinking about a complex collective outcome. We must pursue the cure not by attempting to eradicate cancer or render human nature invincible to it, but rather by sapping it of its power. The only foreseeable species-level cure is not medicinal; it is psychological.

References

Agar, Nicholas. 2015. *The Sceptical Optimist: Why Technology Isn't the Answer to Everything.* Oxford University Press.

American Cancer Society (ACS). 2022. *Lifetime Risk of Developing or Dying From Cancer.* 12 May. https://www.cancer.org/cancer/cancer-basics/lifetime-probability-of-developing-or-dying-from-cancer.html. Accessed 21 Nov.

Banting, Frederick G. 1925. Diabetes and Insulin. *Nobel Lecture.*, 15 September. https://www.nobelprize.org/prizes/medicine/1923/banting/lecture/. Accessed 4 Apr 2020.

Beard, Mary. 2010. *Pompeii: The Life of a Roman Town.* Profile Books Kindle Edition.

Bliss, Michael. 1982. *The Discovery of Insulin.* Toronto: McClelland and Stewart.

Brickman, P., D. Coates, and R. Janoff-Bulman. 1978. Lottery Winners and Accident Victims: Is Happiness Relative? *Journal of Personality and Social Psychology* 36 (8): 917–927. https://doi.org/10.1037/0022-3514.36.8.917.

Deutsch, David. 2011. *The Beginning of Infinity: Explanations That Transform the World.* Penguin Books.

DeVita, Vincent T., and Elizabeth DeVita-Raeburn. 2016. *The Death of Cancer: After Fifty Years on the Front Lines of Medicine, a Pioneering Oncologist Reveals Why the War on Cancer Is Winnable – and How We Can Get There.* Sarah Crichton Books.

Diener, Ed, Lucas Lucas, and Christie Scollon. 2006. Beyond the Hedonic Treadmill: Revising the Adaptation Theory of Well-being. *American Psychologist* 61 (2006): 305–314. https://doi.org/10.1037/0003-066X.61.4.305.

Friend, Tad. 2017. Silicon Valley's Quest to Live Forever. *The New Yorker.* 27 March.

Griffin, Andrew. 2020. AI Solves 50-Year-Old Science Problem in "Stunning Advance" That Could Dramatically Change How We Fight Diseases, Researchers Say. *The Independent.* 30 November.

Harding, Michael C., Chantel D. Sloan, Ray M. Merrill, Tiffany M. Harding, Brian J. Thacker, and Evan L. Thacker. 2018. Transitions from Heart Disease to Cancer as the Leading Cause of Death in US States, 1999–2016. *Preventing Chronic Disease* 15: E158. https://doi.org/10.5888/pcd15.180151.

Hillis, W. Daniel. 2010. On "Cancering.". *Edge.* 27 December.

Hochman, David. 2016. The Playboy Interview With Ray Kurzweil. *Playboy.* 19 April.

Kreizman, Maris. 2019. Why I Am Stockpiling Insulin in My Fridge. *The New York Times.* 9 April.

Lancione, Aly. 2020. 'Flame of Hope' reignited at London's Banting House. *CBC News*, August 27, 2020. Retrieved November 1, 2023 from: https://www.cbc.ca/news/canada/london/banting-house-flame-of-hope-glowing-again-1.5701836

McDonald, Jessica. 2019. Unpacking Biden's and Trump's Big Cancer Promises. FactCheck.org, 19 July. https://www.factcheck.org/2019/07/unpacking-bidens-and-trumps-big-cancer-promises/. Accessed 15 June 2020.

Mossman, Kenneth L. 2014. *The Complexity Paradox: The More Answers We Find, the More Questions We Have.* Oxford University Press.

Mukherjee, Siddhartha. 2011. *The Emperor of All Maladies: A Biography of Cancer.* Scribner.

National Cancer Institute (NCI). 2022. *National Cancer Act of 1971.* https://dtp.cancer.gov/timeline/flash/milestones/M4_Nixon.htm. Accessed 21 Nov.

National Institutes of Health (NIH). 2020. *The Mary Lasker Papers: Cancer Wars.* U.S. National Library of Medicine. https://profiles.nlm.nih.gov/spotlight/tl/feature/cancer. Accessed 7 July.

Niiniluoto, Ilkka. 2019. Scientific Progress. *The Stanford Encyclopedia of Philosophy*,. 16 October. https://plato.stanford.edu/archives/win2019/entries/scientific-progress. Accessed 10 Mar 2020.

Plutynski, Anya. 2018. *Explaining Cancer: Finding Order in Disorder.* Oxford University Press.

Shapiro, James. 2002. Eighty Years After Insulin: Parallels With Modern Islet Transplantation. *Canadian Medical Association Journal* 167 (12): 1398–1400. 10 December.

Sharot, Tali. 2011a. The Optimism Bias. *Current Biology* 21 (23) 6 December 2011: R941–R945. https://doi.org/10.1016/j.cub.2011.10.030.

———. 2011b. *The Optimism Bias: A Tour of the Irrationally Positive Brain*. Random House.

Sontag, Susan. 1978. Illness as Metaphor. In *Sontag Essays of the 1960s & 70s*. Library of America.

World Health Organization (WHO). 2022. Smallpox. https://www.who.int/health-topics/smallpox. Accessed 21 Nov.

Chapter 5
Onward, to Mars!

Contents

While the horizon bias is endemic to our search for decisive technological solutions to problems like climate change, cancer, and aging, it reaches its apotheosis in the aspiration to colonize Mars. Those who are committed to this quest are fully convinced that they can see each step that will be needed to reach their destination.

Hence, in 2015, the astronomer Chris Impey predicted that we will have "small but viable colonies" on Mars within 30 years, while the journalist Stephen L. Petranek anticipated that we will have them within 20 years. Ironically, Petranek's confidence comes despite his apparent awareness of the horizon bias. "Someone drilling for water," he notes, "cannot discover halfway through the process that they have failed to anticipate a specific problem – a mineral deposit that requires a special drill bit, for instance" (Kolbert 2015, no page).

Multiple tech billionaires have stepped up to confront the Mars challenge on humanity's behalf. Chief among them is Elon Musk, who founded SpaceX (Space Exploration Technologies Corp.) in 2002 with the goal of reducing space transportation costs. Though he was inspired by science fiction, particularly Isaac Asimov's "Foundation" novels, Musk presents himself as a true believer in the feasibility of his mission.

In September 2016, SpaceX offered a clear and captivating vision for humanity's future as a spacefaring species. In a short video introducing the "SpaceX Interplanetary Transport System," we can watch a dramatization of settlers launching off from Cape Canaveral and being serenely ferried to a terraformed Mars, complete with oceans and clouds (SpaceX 2016). It looks glorious. The video presents something that Musk expects to happen before the end of this century. He expects that the Mars these travelers land on will have a city of one million people, requiring a fleet of 1000 spaceships. Once established on Mars, humanity will be a multi-planetary species, ensuring that no mishap or misadventure on Earth will send us extinct.

Musk certainly isn't the first person to present alluring visions of our future in space. In 1930, the British politician Frederick Edwin Smith (the Earl of Birkenhead) published a utopian tract predicting that:

> By 2030 the first preparations for the first attempt to reach Mars may perhaps be under consideration. The hardy individuals who form the personnel of the expedition will be sent forth in a machine propelled like a rocket; and equipped with a number of light masts which can be quickly extended, like fishing rods, from its nose. The purpose of these will be to break the impact with which, granted all possible skill and luck, the projectile would strike the surface of the planet. (Birkenhead 1930, p. 132–133)

More recently, the National Aeronautics and Space Administration (NASA) has sought to inspire us with fantastical visions of space-faring humans, not least because its funding depends on the popular public appeal of space travel. Nowadays, NASA "looks forward to writing the next chapter of human spaceflight with its commercial and international partners, advancing research and technology on the International Space Station, opening low-earth orbit to U.S. industry and pushing the frontiers of deep space even further" (Bednar 2014).

Such statements communicate a clear vision of a future in space. But there is an important distinction to be made between NASA's abstract aspirations and the more concrete vision being pursued by Musk. Musk is not content with "sometime in the future." He openly commits to dates and timelines, and he is confident that he will get us to Mars soon.

Musk's confident forecasts are of course riddled with symptoms of the horizon bias. What he envisions is even more extreme than the promise of a prompt cure for cancer or fix for climate change. But that is because he must offer precise dates and specific timelines to stand apart from all those making vague "someday" pitches. The specificity of Musk's plan is a key feature of his sales pitch. It suggests that he already has everything figured out.

There is no doubt that Musk is a talented technologist and entrepreneur. His electric-vehicle company, Tesla, was leaps and bounds ahead of the competition in that sector for at least a decade. Still, whatever he has accomplished there (and previously as a co-founder of PayPal) is dwarfed by the sheer grandeur of his Mars vision. In all of what follows, the reader should bear in mind journalist Shannon Stirone's summary of the hellish conditions on Mars:

> Mars has a very thin atmosphere; it has no magnetic field to help protect its surface from radiation from the sun or galactic cosmic rays; it has no breathable air and the average surface temperature is a deadly 80 degrees below zero. Musk thinks that Mars is like Earth? For humans to live there in any capacity they would need to build tunnels and live underground, and what is not enticing about living in a tunnel lined with SAD lamps and trying to grow lettuce with UV lights? So long to deep breaths outside and walks without the security of a bulky spacesuit, knowing that if you're out on an extravehicular activity and something happens, you've got an excruciatingly painful 60-second death waiting for you. Granted, walking around on Mars would be a life-changing, amazing, profound experience.

But visiting as a proof of technology or to expand the frontier of human possibility is very different from living there. It is not in the realm of hospitable to humans. Mars will kill you. (Stirone 2021)

The Dream Economy

It is Musk's unbridled confidence that helps to explain why he is now one of the world's richest people. Musk's ascent up the billionaire leaderboard owes much to his command of the *dream economy*, which commoditizes thrilling visions of the future. In the dream economy, whether you can marshal persuasive evidence for the feasibility of your vision matters less than your status as a technological vision-ary—the kind of person who can embody our hopes for the future.

In the dream economy, Musk is a technological visionary in the way that Nelson Mandela was a moral visionary in the 1990s. Musk's status as one of the world's richest people suggests we—or at least the markets—are prepared to place our money where Musk's mouth is. Adding to the economic heft of Musk's vision is the fact that even NASA, too, is looking to him to carry out the broader mission.

As a publicly funded agency, NASA must continually justify its existence to politicians. In the excitement following the *Apollo 11* moon landings, politicians were eager to be seen supporting space exploration. In the third decade of the twenty-first century, massive public investments in space travel have become a harder sell. NASA's publicists will still tell you about the importance of the Space Shuttle program, which was terminated in 2011. But now that the agency is confined mostly to flying big planes into Low Earth Orbit, it is no longer meeting the expecta-tions of the generation who watched Neil Armstrong stroll the Sea of Tranquility.

Weren't we supposed to be building habitats on Pluto and tackling the challenges of interstellar travel by now? The feelings of excitement generated by the 1969 *Apollo 11* landing have almost been forgotten. Many of those who look back on the history of space exploration now feel intense disappointment about the apparent lack of progress since 1969. "Had the pace of advances in aerospace travel and technology been sustained in the half-century following the first Moon landing," the astrophysicist Martin Rees observes, "there almost certainly would be human foot-prints on Mars by now" (Rees 2019).

This disappointment is partly attributable to the "low-hanging fruit" problem. To an untutored eye in 1960, stepping on to the lunar surface probably looked every bit as challenging as stepping on to the Martian surface: both terrestrial environs differ greatly from those on which our species evolved. But while the 384,400 kilometers between the earth and the moon takes about 3 days to traverse, the 315.48 million kilometers between earth and Mars will take 6–8 months to cover; count in the return trip, and that is a long time to keep astronauts alive and sane.

Musk thus comes as a welcome salve to an agency that has failed to satisfy the expectations generated by Neil Armstrong's moonwalk. Yet we believe there are significant costs associated with the reflected glare of Musk's charisma. As NASA surrenders more of the future of space to companies like SpaceX, it will likely become a less credible source of information about the practicalities of space exploration. Right now, NASA's public pronouncements are refreshingly free of horizon bias: they point to some of the possibilities of a future in space, but they refrain from offering hard deadlines. One wonders how long this empirically based moderation can withstand Musk's chutzpah.

Another risk of handing things over to Musk is even more obvious: visionaries can mess up. Consider Musk's May 2021 appearance on the US late-night comedy sketch show *Saturday Night Live*, where he took the opportunity to offer a confusing plug for Dogecoin, a cryptocurrency in which he had a stake. When asked, "What is Dogecoin?," Musk answered that, "It's the future of currency. It's an unstoppable financial vehicle that's going to take over the world." But when a cast member responded: "So, it's a hustle?," Musk laughed and said, "Yeah, it's a hustle" (Bentley and Chavez-Dreyfuss 2021).

Musk's stilted delivery left viewers wondering which statement to take seriously. One of his statements must have been a joke, but which one? Since we grant visionaries a special license to calibrate our own expectations for the future, it is dangerous for a visionary to joke. A visionary like Jesus can say, "I am the way and the truth and the life. No one comes to the Father except through me." But he can't then follow that up with a, "Just kidding!" Musk's apparent lack of earnestness about Dogecoin caused its value to plummet 30% while he was still on air (Bentley and Chavez-Dreyfuss 2021).

Cryptocurrencies are different from space exploration, of course. Advocates of cryptocurrencies know that while they can't fool everyone all of the time, it is sufficient to fool some people much of the time. Dogecoin becomes valuable when enough people suspend disbelief and invest in it. Conquering space is fundamentally different. The mere act of suspending disbelief in Musk's plans to colonize Mars won't send humans to Mars. While sufficient collective belief could boost the value of Dogecoin by a factor of a hundred, no amount of collective belief will safely ferry a million colonists to the red planet.

Some contemporary commentators have taken an interest in a passage from the legendary German rocket scientist Wernher von Braun's 1952 work of speculative fiction, *Mars Project: A Technical Tale (Das Marsprojekt)*. There, von Braun imagines a future Martian government "directed by ten men, the leader of whom was elected by universal suffrage for five years and entitled 'Elon'" (Muzaffar 2021).

Obviously, von Braun could not have predicted the birth of Elon Musk in 1971. But this coincidence is marketing gold for Musk and SpaceX. One can only wonder if there is some spooky connection between today's rocket tycoon and the genius whose V2 bombs inaugurated the rocket age in WWII. If so, it doesn't necessarily bode well for Musk. After all, Hitler's hopes that von Braun might turn the tide of the war ultimately were disappointed.

"I Have a Bridge to Sell You"

How, then, should we interpret Musk's Martian forecasts? In this era of "fake news," we contend that SpaceX is offering *fake promises*. When you make a fake promise, you undertake to do something in the future that you don't actually know how to do in the present. If a beggar announces that he is a billionaire, he is lying (fake news). But suppose instead that he undertakes to give you a billion dollars in a week's time. That would be a fake promise. There is no rational basis for trusting that he can get that sum of money together within a week.

One twist in Musk's case is that his wealth and access to capital *almost* imbue his fake promises about Mars with the air of self-fulfilling prophesies. Almost, but not quite. Suppose your local handyman publicly commits to build the world's new tallest skyscraper, exceeding the height of Dubai's 829-meter Burj Khalifa by 10 m. It's unlikely that anyone will pay any attention to him. He has zero experience in building big buildings and thus has issued a manifestly fake promise. But suppose his promise also came with Muskian charisma and the capacity to generate near-unlimited capital. That could make all the difference, allowing him to fake it until he makes it.

Now suppose that people read about the handyman's commitment and draw inspiration from the sheer ballsy-ness of it. He might start to amass supporters who would front him money through GoFundMe. Local news coverage might lead to national coverage, allowing him to market himself through interviews on CNN or Fox News.

As his larger-than-life persona gains purchase, something that was absurd might suddenly start to seem like something he can do. At no point in this marketing campaign did the handyman need to learn anything about engineering or architecture. But that doesn't matter, because, surely, he can find relevantly skilled professionals and can pay them handsomely to help him achieve his—and his investors'—objective. The only missing ingredient is money.

From this perspective, the handyman's confident statements about being able to construct the world's tallest building may contain many engineering and architectural falsehoods and absurdities. But as long as these fibs and exaggerations generate the money to realize the dream, they have done their job. We should view Musk's bold pronouncements about Mars in the same way. They've done their job if they generate the capital SpaceX needs to build sufficiently cool rockets, thereby keeping Musk's Martian dreams alive.

Again, the twist in Musk's case is that the barriers to colonizing Mars are far greater than those blocking the handyman's construction project. The expertise to build an 839-meter building is certainly out there. The handyman and his investors could start by searching out the same team of engineers and architects who completed the Burj Khalifa in 2010.

To mobilize capital and public excitement, it is important for Musk and his marketing department to suggest that Mars will be inhabited not by our long-distant descendants but by some of us living today. Here, we can see how the effort to overcome aging dovetails with the quest to conquer Mars. Not surprisingly, Musk is among those superrich who have already invested in rejuvenation technologies geared toward extending one's lifespan. If these arrive as quickly as those promoting them have promised, investors in their forties could hope to live to see the end of the century, when Mars supposedly will have been terraformed (according to the aforementioned SpaceX video).

Adding to the persuasive power of this imagined timeline is the widely held belief that exponentially advancing technologies will make outcomes that seem far off arrive much sooner than expected. As we'll see in the next chapter, this belief is actively fostered by those hyping new technologies; but it is born of a fundamental misunderstanding.

The horizon bias that makes colonizing Mars seem so imminently doable is an extremely challenging barrier. After accounting for it, we should recognize that Musk is more fabulist than futurist. When the Walt Disney Company wants to devise or implement a creative new idea or technology, it deploys "imagineers" to come up with ideas that are eminently achievable within a specific timeframe. These are the people who create the illusion of traveling to Mars in a theme park ride. If Musk were an imagineer, he would be fired for wasting the imagineering department's time and money with proposals that cannot possibly be finished on deadline.

Cloudy With a Chance of BS

What exactly has Musk proposed? In May 2012, he told PBS *NewsHour*, "I'm talking about sending ultimately tens of thousands, eventually millions of people to Mars and then going out there and exploring the stars" (Boyle 2012). Then in a 2014 interview with *Aeon*, he noted that, "SpaceX is only 12 years old now. Between now and 2040, the company's lifespan will have tripled. If we have linear improvement in technology, as opposed to logarithmic, then we should have a significant base on Mars, perhaps with thousands or tens of thousands of people" (Andersen 2014).

Then, in 2016, Musk suggested that SpaceX's first Mars-shot would be in 2022. By 2020, however, this had slipped back to 2024 for an uncrewed flight, and to 2026 for a manned trip (Cao 2020). In May of 2021, SpaceX launched and then successfully landed a prototype of its Starship spacecraft, and then announced plans to stage its first orbital launch test sometime in the next year (Roulette 2021). Come late 2023, Musk was suggesting that an uncrewed flight might land on Mars before the end of 2027. Musk and his SpaceX staff haven't been sitting on their hands in the meantime, though (Chang 2023). Work on the large Starship rocket that might play an instrumental role in Musk's Mars-shot is progressing at pace. In April of 2023 the first test flight of Starship saw the rocket spin out of control so badly that it was purposefully blown up to avoid the risk of a dangerous crash landing (Chang 2023).

Since SpaceX has indeed chalked up genuine successes, why shouldn't we believe Musk's loftier statements about Mars? The question comes down to what our rational expectations should be. There is no logical principle or law of physics foreclosing on the possibility that some lone genius might announce tomorrow that she has come up with "the cure" for cancer, but we have every reason to discount the likelihood of that scenario.

Similarly, no logical principle or law of physics prevents Musk from being ready to launch a fleet of colony ships ferrying settlers to Mars in 2030. But we wouldn't bet on it.

We have arrived at this skeptical position by applying insights from the work of Philip Tetlock, a psychologist and management professor who has spent his career studying the ingredients that make for credible forecasts. "Foresight isn't a mysterious gift bestowed at birth," Tetlock and Dan Gardner write in *Superforecasting: The Art and Science of Prediction*. "It is the product of particular ways of thinking, of gathering information, of updating beliefs. These habits of thought can be learned and cultivated by any intelligent, thoughtful, determined person" (Tetlock and Gardner 2015, p. 18).

Such habits are clearly distinct from the dress-to-impress forecasting that features prominently on the media. Confident and flashy predictions receive a great deal of press coverage. But as Tetlock and Gardner point out, the best forecasters are "less confident, less likely to say something is 'certain' or 'impossible,' and are likelier to settle on shades of 'maybe.' And their stories are complex, full of 'howevers' and 'on the other hands,' because they look at problems one way, then another, and another. This aggregation of many perspectives is bad TV. But it's good forecasting. Indeed, it's essential" (Tetlock and Gardner 2015, p. 72).

Good forecasters constantly update their beliefs in response to fresh evidence. They make granular assessments frequently, offering credences (degrees of belief) between one and zero—say, 0.2 or 0.3. Upon reading a news story about unrest in Palestine, a good forecaster might shift an earlier credence for peace in the Middle East down from 0.3 to 0.25.

Good forecasters also understand the difference between what can and what cannot be predicted. "Unpredictability and predictability coexist uneasily in the intricately interlocking systems that make up our bodies, our societies, and the cosmos," Tetlock and Gardner explain (Tetlock and Gardner 2015, p. 13). Good forecasters are also aware of the limits on our capacity to make accurate predictions in the first place. With complex systems that involve human beings, the expert predictor's advantage over an amateur essentially evaporates when the time horizon extends beyond 1 year or so.

As we've seen, Musk's genius lies in dress-to-impress forecasting. Another exponent of this approach is the *New York Times* columnist Thomas Friedman, who has long parlayed attention-grabbing soundbites into frequent guest spots on CNN. But Musk's talent clearly exceeds Friedman's. The latter has earned many guest spots on TV, but Musk's pronouncements have made him a multi-billionaire.

However, Musk does not seem to be emulating the techniques of what Tetlock calls superforecasters—those who outcompete successful "dress-to-impress" forecasters by constantly revising their credences. Instead, Musk conforms to the pattern of the visionary who has found the Truth. For him, each encounter with fresh evidence inevitably deepens his confidence in himself.

To be sure, there are certainly circumstances under which it would be rational for a forecaster to increase his confidence in our ability to establish a self-sustaining colony on Mars. Demonstrable and measurable progress toward the red planet would allow us to strengthen our credence. Suppose that SpaceX had persisted and succeeded with "Mars Oasis," its plan in the early 2000s to land a small greenhouse on the planet to begin the propagation of crops.

By 2021, there would be a tiny garden of plants on Mars contributing a tiny amount of oxygen to the Martian atmosphere. In this alternative timeline, it would be rational to increase (very slightly) the odds of settling humans on Mars, because there would have been a proof-of-concept for terraforming the planet.

Musk certainly does have some cool rockets, and he has made progress toward making them reusable—an absolute must for any plan to send a million people to Mars before the century is out. But when we examine Musk's statements and their growing confidence over time, we see the work not of a rational forecaster but of a dreamer who is determined to fake it until he makes it.

To be sure, Musk is certainly not the only one marketing a future in space. Though it is more prudent about setting deadlines, NASA's own sales pitch is equally aspirational. And Musk is also joined by other billionaires like Jeff Bezos and Richard Branson. In July 2021, the latter "made history" by becoming the first private citizen to launch himself into space (something the Soviets did 60 years ago when they sent cosmonaut Yuri Gagarin into orbit for an hour and 48 min) (Norris et al. 2021).

Bezos and Branson's space proposals seem more credible than Musk's. One reason for this is that Musk has already staked a claim to the idea of colonizing Mars. There are diminishing reputational returns from saying "Just like Elon, I'm going to terraform Mars and settle humans on it." What about going to Venus? That prospect leaves other billionaires with less enticing marketing opportunities.

Moreover, the name of Bezos's space company, Blue Origin, hints at the interest that he claims originally propelled him into space. He wants to move high-pollution heavy industries into space so as to preserve the habitability of Earth. That kind of pitch certainly has appeal in an age of ecological anxiety. With heavy industry shifted off-world, we can effectively have our cake and eat it too. All the consumer durables we could ever want would be ferried down to us while the pollution generated by their creation wafts off into deep space.

Still, from a marketing perspective, this is inferior to Musk's pitch for Mars. Heavy-industry space stations feel much more dystopian than utopian. Why clutter up space we can create a whole new Earth on Mars?

Prediction as Propaganda

Ultimately, Musk fits the description that Henry David Thoreau applied to a proto-Muskian hype-man of his own time: "He has more of the practical than usually belongs to so bold a schemer, so resolute a dreamer. Yet his success is in theory, and not in practice, and he feeds our faith rather than contents our understanding" (Thoreau 1843, p. 462).

In his Mars vision, Musk is promising something that still lies over the horizon. Armed with his manifesto science, he has compared himself to Daedalus, the great technologist of Greek mythology. Others have duly likened him to Icarus, Daedalus's son, who flew too close to the sun (Oremus 2021). But, as we've seen, Musk's overly confident predictions reflect method, not madness. For him, hype serves a useful purpose that goes beyond merely generating free press (though it certainly does do that). Like those forecasting a millennial lifespan or an AI singularity, Musk is in the mythology business. As the *fin de siècle* French theorist Georges Sorel observed, such secular myths exploit the unpredictability of the future by "framing" it in ways that are useful for "acting on the present" (Sorel 1908, 114–116). For someone in Musk's position, a vague, indeterminate expectation that humanity will colonize Mars is a lucrative thing.

Better yet, as an expression "of a will to act," myths, Sorel notes, bring "few inconveniences." In issuing bold predictions, Musk has little to lose and much to gain. In the dream economy, the content of the promise is secondary to the attitude it instills:

> A knowledge of what the myths contain in the way of details which will actually form part of the history of the future is then of small importance; they are not astrological almanacs; it is even possible that nothing which they contain will come to pass – as was the case with the catastrophe [the Rapture] expected by the first Christians. (Sorel 1908, 116)

In this way, the act of prediction becomes an act of propaganda.

As a futurist-propagandist, Musk and other contemporary techno-utopians are part of a long tradition. Since Thomas More's *Utopia* in 1516, I.F. Clarke reminds us, the future, as an idea, has been contested. Detailed *fictional* descriptions of what lies over the horizon serve as roadmaps for potentially effecting change in the real world.

Those who can articulate a compelling prophesy are not merely making a guess about the future. They are actively determining which aspirations will "define the direction of future developments." Hence, Simon Lake, the American engineer who invented one of the first submarines, revealed in his memoir that he took his inspiration from the father of science-fiction, Jules Verne (Clarke 1979, p. 4–8, 201).

If we bear in mind that most predictions are propaganda that is acting on the present, we will be better placed to sort between hype and sound forecasting. The task, for all of us, is not just to determine whether there is a strong rational basis for believing what the exponents of Silicon Valley are promising us. We also must explore and keep alive the promising alternative imaginaries have been crowded out by the dreams of a few billionaires. Instead of aspiring to render Mars inhabitable, we could focus on ensuring that Earth remains so.

Maintaining such a perspective is all the more important in an age where "exponential" technological progress will continue to expand the scope of the possible— the topic we turn to next.

References

Andersen, Ross. 2014. Exodus. *Aeon*. 30 September.

Bednar, Scott. 2014. NASA 360 – The Future of Human Space Exploration. NASA.gov. 4 November. Last updated 7 August 2017. https://www.nasa.gov/content/nasa-360-the-future-of-human-space-exploration-trailer. Accessed 7 July 2021.

Bentley, Alden, and Gertrude Chavez-Dreyfuss. 2021. Dogecoin Tumbles After Elon Musk Calls It a "Hustle" on "SNL" Show. *Reuters*. 7 May.

Birkenhead. 1930. *The World in 2030 A.D.* Brewer and Warren Inc.

Boyle, Alan. 2012. SpaceX Chief Wants to Be Spaceflier. *NBC News*. 3 May.

Cao, Sissi. 2020. Elon Musk Reveals SpaceX's Timeline for Landing Humans on Mars. *Observer*. 2 December.

Chang, Kenneth. 2023. Elon Musk Says SpaceX Could Land on Mars in 3 to 4 Years. *The New York Times*, 5 October. Retrieved November 2, 2023 from: https://www.nytimes.com/2023/10/05/science/elon-musk-spacex-starship-mars.html

Clarke, I.F. 1979. *The Pattern of Expectation, 1644–2001*. New York: Basic Books.

Kolbert, Elizabeth. 2015. Project Exodus. *The New Yorker*. 1 June.

Muzaffar, Maroosha. 2021. In 1953, a Scientist Said the Title of the Leader of a Martian Government Would Be "Elon.". *The Independent*. 8 July.

Norris, Kelcey, Susan Montoya Bryan, and Marcia Dunn. 2021. *Billionaire Richard Branson Makes History, Reaches Space in His Own Ship*. Associated Press. 11 July.

Oremus, Will. 2021. The Two Sides of Elon Musk. *The Washington Post*. 16 July.

Rees, Martin. 2019. Are Moonshots Still Possible? *Project Syndicate*. 19 July.

Roulette, Joey. 2021. From Texas to Hawaii: SpaceX Plans First Orbital Starship Test. *The Verge*. 13 May.

Sorel, Georges. 1908/1999. *Reflections on Violence*. Cambridge University Press.

SpaceX. 2016. SpaceX Interplanetary Transport System. *YouTube*. 27 September. https://www.youtube.com/watch?v=0qo78R_yYFA. Accessed 7 July 2021.

Stirone, Shannon. 2021. Mars Is a Hellhole. *The Atlantic*. 26 February.

Tetlock, Philip, and Dan Gardner. 2015. *Superforecasting: The Art and Science of Prediction*. Crown Publishers.

Thoreau, Henry David. 1843. Paradise (to Be) Regained. *United States Magazine and Democratic Review* XIII (XLV): 451–463.

Chapter 6
But, What About Exponential Progress?

Contents

As we've seen, an unjustified degree of optimism about the future of disease, space travel, and other challenges afflicts not just hype-men with something to sell, but also many of the best-informed scientists. Those with enough knowledge to be able to rule out obvious *impossibilities* often are lured toward the Siren's call of "how-possibly" plausibility. The horizon bias—born of the modern faith that the right knowledge and technology will quickly resolve all problems—perennially obscures obstacles standing in our way.

But if we are forced to make concrete predictions about cancer cures or the colonization of Mars, we would have powerful inductive reasons to temper our optimism. We should expect some intractable hitch that prevents the theory from being put into practice. There is no good reason to believe that the universe was designed to satisfy the rosiest expectations of researchers and technologists circa 2020.

But what about all the recent progress in private space flight and "health-tech"? One can scarcely go a day without reading headlines about machine learning and Big Data being brought to bear on previously intractable medical problems. AI pattern-recognition and deep-learning methods are reported to have already surpassed human radiologists in classifying skin cancer (Hosny et al. 2018, p. 500). The physician and best-selling author Eric Topol believes that these technologies can be unleashed to render a medical phenotype of every individual, diagnose a growing range of diseases, and discover new drugs (Topol 2019).

Going much further, the futurist Byron Reese (who is not a physician) argues in *Infinite Progress* that, "Armed with the data to develop medical knowledge and wisdom, and the technological tools to enable medical progress at an ever-quickening pace, we can confidently foresee a day when humanity will overcome" *all* diseases (Reese 2013, p. 72). His confidence is based on the exponential growth of computing power, which allows for the ever-faster analysis of ever-more data.

Reese asks us to, "Imagine a computer culling through" the "inconceivably large" set of existing data on skin cancer. "Pulling out patterns," it might find that, "People who eat radishes get better slightly more frequently than people who don't," which in turn might lead the machine to find that certain pesticides used on radishes "might just cure skin cancer" in people who have a "certain common, though not universal, genetic market." Reprising the hopeful speculation of the late Clinton era, when we welcomed the first draft sequencing of the human genome, Reese wants us to consider, "How many connections like [this] there are in the universe – causes and effects we cannot see because the sheer amount of data we encounter in the everyday task of living is overwhelming our minds" (Reese 2013, p. 72–74).

Echoing Kurzweil, the high priest of exponential progress, Reese concludes that "We will do much more in the next twenty years than in the preceding one hundred. After that, more in five years than those twenty. Then more in one year than those five. Given all this, do you really believe disease has a chance?" (Reese 2013, p. 74).

Such enthusiasm is *de rigeur* in the tech world. Recall Kurzweil's prediction that within 10–12 years, we would have reached a point where we had a cure to cancer either in hand or within sight (Diamandis 2017). Like Reese, his reasoning was based not on any particular cancer-fighting strategy, but on the pace of technological innovation in itself. The cure for cancer will follow naturally from the uptake of biotechnology, which turns all biology into information processing, which itself is advancing exponentially.

But even if one accepts that "technological progress" is accelerating, that doesn't mean a cure for cancer is imminent. The argument assumes incorrectly that we already know what it would take to cure cancer, even though we will not have that knowledge until we've already done it. Short of that information, claims about exponential technological progress aren't nearly as meaningful as they sound.

Obviously, if something that counts as a cure is indeed achievable, then *any* progress will bring us closer to it, and accelerating that progress will shorten the duration of the process. But unless we know the distance to the destination, we cannot know when we will arrive. To the children in the backseat chanting "Are we there yet?," the minute-by-minute experience of riding in an accelerating vehicle will feel the same regardless of whether the destination is three or 30,000 miles away. Parents, of course, tend to limit their road trips to single-day affairs. But in the case of the War on Cancer or the colonization of Mars, that decision isn't up to us.

Why This Time Isn't Different

Just as we rely on metaphors to organize our engagement with the Big Cs in lieu of a solution, so do we think figuratively about concepts like exponential progress—whether or not we realize it. Here, as in the previous cases, metaphor can prove either helpful or harmful depending on how we use it. The risk, as always, is that the metaphor can become "realized," defining rather than merely assisting our thinking about the problem. Unfortunately, this has certainly been the case with techno-optimist accounts of acceleration.

Consider, for example, the notions of "inflection point" and "critical mass," two terms that one encounters all too often in arguments touting the pace of growth in any given domain. In December 2017, for example, the commissioner of the US Food and Drug Administration, Scott Gottlieb, announced before a US Senate committee that, "We stand at an inflection point in medicine – where new technology is creating foundational opportunities to treat and cure disease in ways that weren't possible just a short time ago." Citing recent, limited successes with CAR-T gene therapy to treat certain cancers, Gottlieb set his sights firmly on the horizon. "This experience shows how a single, fundamental breakthrough in science can open up a whole new way of combatting disease" (Gottlieb 2017).

But the CAR-T saga turned out to be a classic case of over-hype. The treatment is effective in only a small population of patients, and has not yielded anything close to the universal applications initially promised by its proponents (Raza 2019, p. 211–215). Far from an inflection point, it merely reminds us that cancer remains in the category of "high-hanging fruit." That is physician and author James le Fanu's own metaphor for moderating our thinking about medical progress, which he believes has been decelerating since the mid-twentieth century. Insulin, antibiotics, cortisone (steroids), open-heart surgery, preventative and therapeutic measures against some cancers (namely, the smoking connection)—these were among the low-hanging fruit.

The frequency of such "definitive moments" has been in secular decline, because the problems that remain simply do not admit so readily of solutions. This has resulted in deep frustration. Picking the low-hanging fruit created a "common perception" of "medicine's historic achievements … being on a continuous and upward curve of knowledge," le Fanu writes in *The Rise and Fall of Modern Medicine*. "Here the unknown is merely waiting to be known with, in principle, no limits to its further beneficent advance" (Le Fanu, 2011, p. xviii).

When it comes to technological fixes for cancer, the future can be characterized in relation to either of two curves: le Fanu's, which projects diminishing returns; or Kurzweil's, which foresees accelerating progress. Le Fanu's fruit-tree argument implies that the steepening of the curve toward ever higher planes of medical achievement is an illusion. Indeed, a plausible explanation for the failure to win the War on Cancer is that we're now left reaching for ever-higher fruit. The fact that we have made major achievements against cancers like Acute Lymphoblastic Leukemia

(ALL) implies not that we are accelerating progress, but rather that those forms of the disease were easier to treat than many others.

The second curve depicts exponential progress and is in competition with the first curve. If accelerating technological progress does translate into exponentially greater power over things like cancer (by way of data-analytics and other corollaries), we should expect to see more, faster progress. But if the first curve exerts a more powerful influence, then significant advances against cancer will take longer to arrive, and the net effect will feel like a deceleration of progress.

ABC, Easy as 1-2-4

Kurzweil, Reese, and other techno-enthusiasts' hope about cancer (and everything else) stems from one of the defining principles of the digital age: Moore's Law. In a 1965 paper, Intel co-founder Gordon Moore observed that the number of transistors on an integrated circuit tended to double about every 2 years (Moore 1965, p.83). For decades, this rate of progress was borne out (though it obviously cannot continue forever), leading Kurzweil to develop his own Law of Accelerating Returns: "the rate of change in a wide variety of evolutionary systems (including but not limited to the growth of technologies) tends to increase exponentially" (Kurzweil 1999, p. 30, 32).

Precisely because his principle comprises all forms of progress, Kurzweil has no problem collapsing human biology into the same category as technology. Our hereditary/genetic material is merely so much information to be processed. And since our information-processing power is expanding exponentially, we should assume that "the speed reflected by our current technologies is nothing compared to what is to come."

As we will see, evangelists of technological progress use the term "exponential" with reckless abandon, usually treating it as a synonym for "explosive," and as a sufficient condition for the imminent achievement of whatever goal they have in mind. By definition, an explosion is an event in which things undergo radical, instantaneous change. Yet it is worth pausing to consider how mathematicians understand exponential change: namely, as an instance of some quantity increasing or decreasing at a rate that is proportional to the quantity itself. In cases of exponential growth, this results in acceleration, because each iteration of increase is greater than earlier iterations.

If it seems implausible that translating biological information into bits of data will rapidly lead to a cure for cancer, Kurzweil would contend that we're still at a relatively early stage in the process, when exponential improvements seem small. In an exponential doubling sequence, an early stage of "1, 2, 4, 8" looks a lot like the early stage of an arithmetic "add 2" sequence: "2, 4, 6, 8." But after an initial period in which the two sequences are barely distinguishable, they start to diverge radically. Suddenly, "10, 12, 14, 16, 18, 20, 22" is put next to "16, 32, 64, 128, 256, 512, 1,024."

A classic demonstration of the principle is the story of the king who offers to pay his wise adviser in rice. The adviser proposes that the king place a single grain on the first square of a chessboard, and then merely double that each day. Spotting a bargain, the king immediately agrees—the man seems to be offering 2 months' worth of service for a mere scoop out of the royal food store. The king hasn't done the math, of course. A chessboard has 64 squares, and 2^{63} (18,446,744,073,709, 551,615) grains of rice is actually far more than the kingdom could ever produce.

This is a powerful demonstration; for the mathematically uninitiated, 18.5 quintillion is simply an unfathomable number. If the apostles of exponential progress are to be believed, the powers of AI and other digital technologies will soon surpass the bounds of what we can even imagine (Agar 2015, Chap. 2). In 1982, the Sinclair ZX Spectrum personal computer was marketed as having up to a "massive 48KB of RAM." Today, RAM (random access memory) in consumer electronics is quantified in gigabytes. Nobody thinks twice about multiples of 100,000,000 bytes, even though these figures make a mockery of the Spectrum's "massive" 48,000 bytes.

Beyond 1980s hardware and fables about wise men and kings, techno-optimists' favorite exhibit is the "hockey-stick" curve: the depiction of exponential growth on an x-y axis. As Americans, in particular, learned at the height of the COVID-19 pandemic, when infections were increasing exponentially, the hockey-stick curve is easy for anyone to understand. After a slow and unspectacular crawl horizontally, the line bends upward and quickly comes to appear vertical, at which point the sky is the limit. Depending on what is being measured, this could be either a very good thing or a very bad thing.

The most iconic exponential-growth curve features a simple doubling sequence, such as in depictions of doubling grains or repeatedly folding a sheet of paper in half (if you could do it 103 times, you would have a wad of pulp thicker than the observable universe—93 billion light-years).

But the tyranny of "times two" oversimplifies our understanding of exponential change. There are many examples of exponential growth that don't seem especially explosive. Suppose you serially multiply a quantity by 1.00001 once per day. This would be exponential growth—the total quantity would increase by a greater amount each day. But if the exponential growth of COVID-19 infections had followed this pattern, it would not have become the problem that it did. There would have been no need for Operation Lightspeed; Operation Snailspeed would have produced the vaccine well before the wider population was at risk of infection.

Exponential-growth curves have long featured in the study of material progress more broadly. In the late 1920s, John Maynard Keynes predicted that the "power of accumulation by compound interest" and ever-rising productivity growth would increase the "standard of life in progressive countries" four- to eightfold over the next 100 years, allowing for a 15-h work week (Keynes 1930). Keynes's prediction about the rate of future growth was considered far too optimistic at the time, given that the rate of growth had fallen after World War I, and that he lacked the data and forecasting tools that are now taken for granted today. But his confidence turned out to be well placed; incomes actually grew far more than he predicted.

Keynes's mistake, of course, was to translate his foresight about exponential economic growth into social and cultural prophecy. He envisioned a fully automated steady-state society in which most people would pursue the "art of life" through leisure. But he ignored the complications posed by population growth, inequality, and the insatiability of human desires and ambitions. Owing to hedonic normalization, each subsequent generation in the rich world has regarded its inherited state of relative abundance not as Keynes's endpoint, but as its own starting point for evermore accumulation.

Moreover, in the advanced economies that Keynes was thinking about (the developing world was another significant lacuna in his vision), there has been much handwringing among economists over the failure of growth to deliver what it once promised. In these debates, the Northwestern University economist Robert J. Gordon offers his own "low-hanging-fruit" argument about productivity-enhancing innovation, contending that today's technologies simply are not as consequential as previous breakthroughs such as electrification, indoor plumbing, the internal-combustion engine, and so forth (Gordon 2016).

Representing the opposing view are MIT economists Erik Brynjolfsson and Andrew McAfee, who point out that it generally takes a number of decades for the full effects of a major technological breakthrough to seep through the broader economy. That was the case with the steam engine, and it will prove true for computers and robotics, too. The past few decades of relatively tepid growth were merely a temporary lull before an explosion of economic dynamism. With the digital era, we have once again reached what Brynjolfsson and McAfee call an "inflection point":

> Self-driving cars, *Jeopardy!* champion supercomputers, and a variety of useful robots have all appeared just in the past few years. And these innovations are not just lab demos; they're showing off their skills and abilities in the messy real world. They contribute to the impression that we're at an inflection point – a bend in the curve where many technologies that used to be found only in science fiction are becoming everyday reality. As many other examples show, this is an accurate impression (Brynjolfsson and McAfee 2014, p. 34–37)

Brynjolfsson and McAfee conclude that, "The digital progress we've seen recently is … just a small indication of what's to come," because "the nature of technological progress in the era of digital hardware, software, and networks … is *exponential, digital,* and *combinatorial.*" Hence, when it comes to fighting cancer, they think it "not implausible that [IBM's question-answering AI] Dr. Watson might one day be the world's best diagnostician" (Brynjolfsson and McAfee 2014, p. 93).

Sounding a similar note, the *New York Times* columnist Thomas Friedman believes that the year 2007 "may be understood in time as one of the greatest technological inflection points" of the modern era. In that year, Google bought YouTube and introduced the Android mobile operating system; IBM launched Watson; Amazon released the Kindle; Satoshi Nakamoto conceived of Bitcoin; AirBnB was founded, and Netflix began its video-streaming service (Dizikes 2018).

Before all of these aforementioned "inflection points," our arithmetic intuitions were adequate for understanding technological change; but after them, the rate of change supposedly transforms in some fundamental way—or so best-selling authors like Friedman, Brynjolfsson, and McAfee would like us to think.

However, there is a glaring problem with these arguments: they all rely on a non-standard—indeed, metaphorical—meaning of inflection point. In mathematics, an actual inflection point is where the *value* of a curve changes between positive and negative, concave or convex.

This is no mere technical distinction. A true inflection point is an objective feature of a trend line, which means that one's perception of it will not change when one zooms in or out on the graph. By contrast, what Friedman, Brynjolfsson, and McAfee describe as an "inflection point" is wholly subjective. The "explosions" of innovation they perceive are purely impressionistic. In a graph depicting exponential improvement, one can designate essentially any point as seeming explosive.

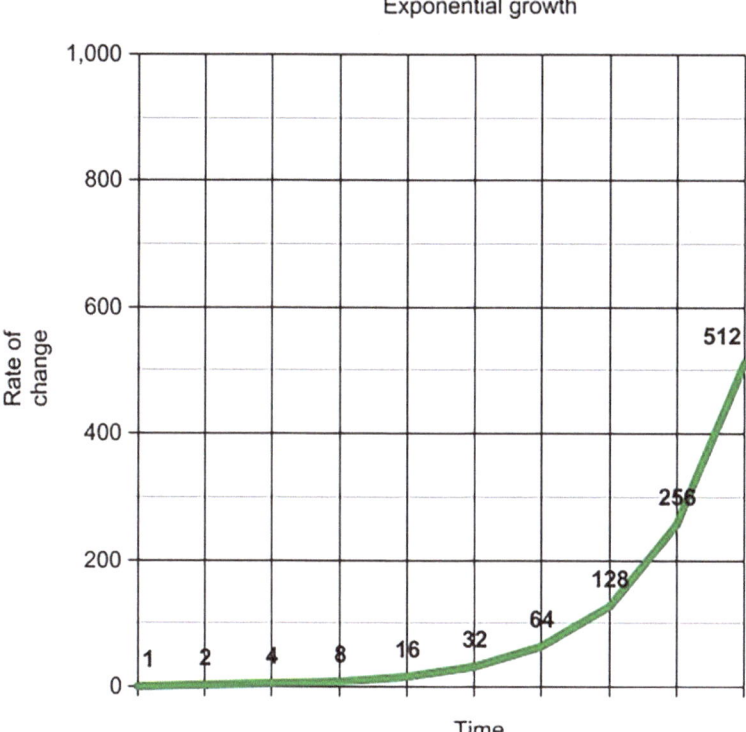

Is the "explosion" at value 128? ...(*Created by authors*)

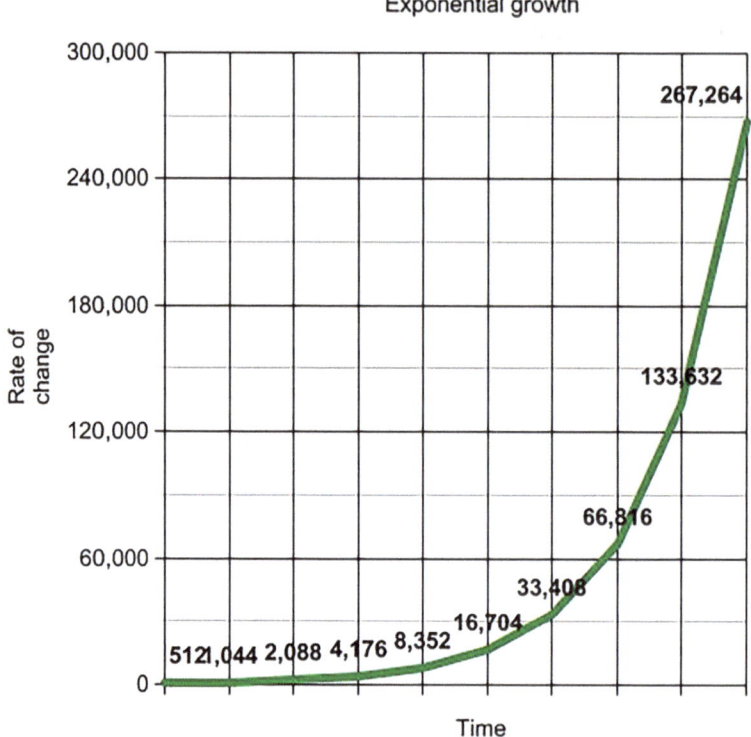

Or is it at value 33,408? (*Created by authors*)

Viewed in this light, we can see that Gottlieb, Brynjolfsson/McAfee, and Friedman's "inflection points" are each in the eye of the beholder. The term is being used metaphorically, as a means of emphasis. A true inflection point would show the curve bending *down*. But in popular accounts of progress, it is used figuratively by those who want us to believe that a specific point in time—a certain development in a longer-running trend—is special in the grand scheme of things.

Without a sense of the destination, however, such assertions are unhelpful, at best. As our past record of confronting cancer shows, many improvements that contemporary observers regard as inflection points or "game-changers" will be mere footnotes in future history books, assuming that they bear mentioning at all. Technological progress that corresponds with victory in the War on Cancer may well be waiting for us at value 534,528. Yet it also could be much further off at, say, 1.37^{19}.

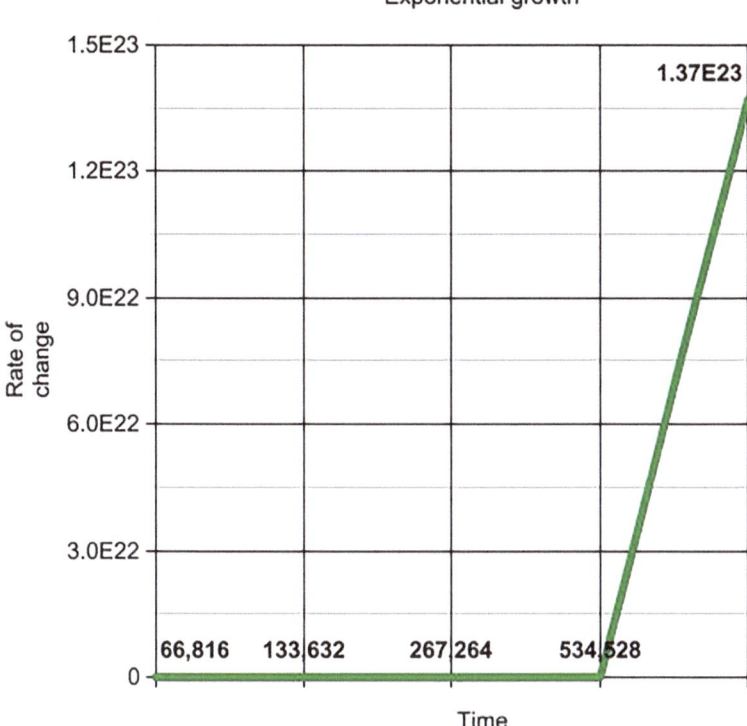

What Value for Victory Day?. (*Created by authors*)

So, even if one accepts that technology is improving exponentially, one should be skeptical of claims that it has reached an "inflection point." But that brings us to another, closely related metaphor that one inevitably encounters in optimistic accounts of progress: the "bend in the knee." This conceptual device suffers from the same problem as that of "inflection points." As the web programmer Toph Tucker explains, "Some things in nature have a knee. Exponential curves are not among them. There is no unambiguous point at which it really starts gettin' going. Anything you point at as the knee is simply a matter of perspective" (Tucker 2016).

For the sake of argument, let's grant that there are indeed observer-relative "inflection points" or "bends in the knee" on an exponential curve. Two observers looking at the same curve could each designate a different "knee," and neither would be wrong. One person's "slow-growth phase" may be another's inflection point, and yet another's "rapid-growth phase." In Brynjolfsson and McAfee's case, the central contention is that we have reached "a point where the curve starts to bend a lot – because of computers" (Brynjolfsson and McAfee 2014, p. 9).

Here, "bend" is an objective property of the curve. But "a lot" is a subjective judgment that assumes a particular perspective. If you double a single donut crumb, you will have two crumbs, or something like 0.0002 donuts. From a human perspective, such increases will be negligible until you've reached the realm of whole donuts. It is generally advisable that one does not eat more than one or two donuts in a sitting; but, let's be honest, if one is hungry enough, one can consume quite a bit more.

Nonetheless, once the number of donuts is being doubled from 25 to 50, then to 100, then to 200, it is not implausible to identify an "inflection point." The shift from double to triple to quadruple digits each could qualify as points "where the curve starts to bend a lot." But change the perspective, and the inflection points will change, too. To an ant, that first act of doubling 0.0001 donuts to 0.0002 would seem more impressive. Indeed, an ant would expect to see an inflection point quite soon as the doubling process starts to yield quantities far in excess of what any ant could consume over many lifetimes. To the ant, the point of "explosive" growth will arrive earlier than it would for the human, and far earlier still than it would for an elephant.

This Changes Everything!...

Another big event in 2007 was Apple co-founder Steve Jobs's announcement unveiling the first-generation iPhone. "In one device," he boasted, "we will have the world's best media player, the world's best telephone, and world's best way to get to the Web – all three in one" (Friedman 2016, Chap. 2). Even if one doubts Friedman's suggestion that Netflix's pivot into video streaming represents a signal event in the history of human progress, one must admit that the arrival of the iPhone the next year was a moment that "changed everything." Right?

Not necessarily. When we have been culturally conditioned to expect constant, consistent, and exponential improvements in technology, we tend to inflate the importance of each new advancement, while discounting the implications of everything that came before it. But the question isn't whether the latest iPhone blows all past generations of handheld consumer electronics out of the water. Obviously, it does. Thanks to Moore's Law, the newer product will be more powerful than previous generations by orders of magnitude. Each iteration is ineluctably unprecedented. Someone who can still remember the pre-smartphone days will be inclined to think that she is witnessing a point of departure: the dividing line between slow linear-seeming exponential improvements and radically explosive improvement.

The problem, of course, is that when each new iteration is unprecedented, an "inflection point" is always imminent. The release of the first iPhone will almost certainly merit a spot in any future history of mobile communication, whether the story is told in 2028 or 2280. But whether this particular consumer electronic represents a game-changer, as opposed to just another advance among many others, depends, once again, on perspective.

In a book published in 2010, the iPhone 1 gets its own chapter; but by 2110 it may have been relegated to an endnote, along with the Blackberry, the Palm Pilot, and other handheld precursors. The differences between Nokia dumbphones and the iPhone were immensely salient to people judging from the year 2008, but they will seem less so with each passing generation. Some future historians might choose to elevate the iPhone and its charismatic creator; but others might find more value in a revisionist perspective, emphasizing the under-appreciated contributions of other technologies.

Indeed, there is a future dissertation to be written about the forgotten historic import of Nokia, a Finnish telecommunications company that was founded in 1865, and whose 9000 Communicator series mobile phones offered (expensive) internet access as early as 1996. By 1999, the Nokia 3210 featured internal antennae, customizable ringtones, three preinstalled games, and T9 text entry in a compact, cheap format that was especially appealing to young professionals.

Whether the first iPhone represents a "knee" on the curve is thus temporally relative. It certainly seemed knee-like from Jobs's vantage point in 2007; but the superlative already seems less obvious in the early 2020s. Millions of earlier-generation iPhones have been discarded and replaced with more recent iterations, each of which tends to be less impressive to users as time goes on. The camera gets better, the screen resolution improves, the battery lasts longer—but there is no longer a subjective sense of novelty, let alone revolutionary potential.

The problems of hedonic adaptation and, particularly, hedonic normalization have implications for where one places one's "inflection point." As a classic in the annals of design, the iPhone 1 doubtless deserves a place alongside the Volkswagen Beetle and the Bic ballpoint pen. But such judgments are aesthetic. If a 2024 iPhone 16 Pro user suddenly returned to the iPhone 1, he would probably find it rather primitive, what with its 2.0-megapixel camera, blurry display, and limited storage. Far from a marvel out of science fiction, it wouldn't feel all that different than an old Blackberry, or even a Nokia 3210.

In thinking about the recent past and future of technology, we should consider another objective feature of exponential curves. Suppose that technological "progress" follows a doubling pattern (1, 2, 4, 8). As such, the most recent iteration at any point in time will not only be a bit more impressive than the one immediately preceding it; it will be twice as impressive. Not only that, its significance will be at least twice that of all previous iterations combined.

After all, if the exponential sequence "1, 2, 4, 8, ..." describes progress in computing power, the most recent iteration will have a power of 8, whereas the sum power of all preceding iterations will be 7. By its very nature, the doubling sequence delivers more than has ever come before. But if that is the case, why should the latest iteration be interpreted as an "explosion"? The same could be said about any single instance in the chain of exponential progress.

In the world described by Kurzweil's Law of Accelerating Returns, the present will always be an exciting time to be alive (and tomorrow will always be even more exciting than today). The disappointment doesn't come until we pause and reflect on the past. That is when we can see that previous "explosions" were anything but. At the end of the day, enthusiastic declarations of "inflection points" are no more than intimations of the speaker's own excitement. The new and novel will always seem special, particularly when we have been primed always to expect exponential improvements and imminent breakthroughs or the new product is promoted by someone in a black turtle-neck.

Going Nuclear (Metaphorically)

An equally potent but misleading metaphor is the notion of "critical mass," which has been taken up (rather ill-advisedly) by some cancer researchers. In physics, "critical mass" refers to the minimum amount of fissile material required to create a nuclear chain reaction. In the case of cancer, advocates speak of a "critical mass" of knowledge, as if to suggest that the only remaining hurdle to a cure is the lack of resources or will power. With an injection of cash, a chain reaction will ensue, leading imminently to new breakthroughs.

According to Vincent DeVita, Mary Lasker, the philanthropist and principal figure behind the launch of the War on Cancer in the 1970s, used this metaphor as a rallying cry against the emperor of all maladies. She genuinely believed that the scientists of her time knew enough about cancer to strike a decisive blow against it. Recently reported successes with chemotherapy had convinced her that the missing piece to the therapeutic puzzle—effective systemic therapy—had been found. But she "was wrong," DeVita writes. "We had not achieved critical mass at that time." Still needed was a critical mass of what he calls "usable knowledge" (DeVita 2002, p. 354).

Writing in 2002, DeVita suspected that we finally had acquired the knowledge and technologies needed to reduce cancer mortality significantly. Then, in a 2015 book (co-authored with his daughter, Elizabeth DeVita Raeburn), he again suggested that "we will see the end of cancer as a major public health issue. … we have the critical mass of knowledge to get us the rest of the way." The remaining obstacles, he argues, "are not scientific":

> Rather, they are in the form of not using what we know and the tools we already have to cure more, because of a reluctance to drop outdated beliefs, bureaucratic battles among physicians and medical groups, and a Food and Drug Administration that has not caught up with the innovations in cancer drug development. (DeVita and DeVita-Raeburn 2016, p. 8)

From a vantage point in the early 2020s, it is clear that a critical mass of knowledge and technologies for winning the War on Cancer in fact did not exist in 1970. Less clear is how observers in 2070 will see things. What will they make of DeVita's claims in 2002 and 2015? The 2002 instance predates many of the significant advances in immune-oncology, and in his 2015 book, DeVita counts immunotherapy among the "breathtaking" new therapies that have convinced him that victory is near (DeVita and DeVita-Raeburn 2016, p. 8).

Will the author of a 2070 retrospective regard DeVita's claims about "useable knowledge" in the same way that DeVita looked back on Lasker's aspirations in 2002 and 2015? From the perspective of 2070, the years 1970, 2002, and 2015 will each correspond to points of genuine progress against cancer. They may even coincide with periods of exponential progress against cancer. But where will the critical mass be placed? Such judgments are as arbitrary as those for "inflection points" and "bends in the knee."

A caveat is in order here. We are discussing rational expectations about humanity's engagement with hard problems. An underestimation of the Big Cs' complexity and an overestimation of our own abilities have been two consistent features of recent history.

While no one can say with certainty that there won't be a breakthrough next year that cures all cancers or solves climate change, there certainly isn't enough evidence to think it's likely that these things will happen. A sudden "miracle" should only be met with complete surprise, not understood as a development that is long overdue. The discovery of an immunotherapy that works on all types of cancer, for example, would imply the surprising finding that cancer is actually much simpler than we understood it to be.

But we also used to think that infectious diseases were simple. After World War II, notes Alex de Waal in *New Politics, Old Pandemics*, "international health campaigns against smallpox and polio registered huge gains. The successes were such that the medical academy was ready to declare victory over infectious diseases altogether." The post-war generation had already forgotten the 1918–1919 influenza pandemic, which had "refuted civilization's proudest achievements," and it could not have foreseen HIV/AIDS (De Waal 2021, p. 12–13).

The HIV/AIDS crisis demonstrated that we still have much to learn about infectious diseases and the immune system. HIV/AIDS defied all attempts to develop a vaccine, and even exhibited properties similar to cancer, not least the ability to develop resistance to drug treatments. As of the early 2020s, HIV/AIDS has not been cured or eradicated. Rather, it has been made into a chronic condition through combination drugs that target different replication and growth mechanisms within the virus simultaneously (Haseltine 2020).

Until we've achieved consistently similar results with all forms of cancer, we simply will not know how complex the disease is relative to our current understanding of it. If cancer turns out to be as simple as Kurzweil assumes, the impressive advances that inevitably will be made over the next decade will bring us close to victory. By contrast, if it turns out to be more complex, those same improvements will still move us forward; but they will leave us far short of the finish line. Cancer specialists in 2070 doubtless will look at *their* most recent advances as game changers, while remembering the progress made in the 2020s as unimpressive. Today's cutting-edge immunotherapies will someday hold the same position as the Nokia 9000.

Unknown Unknowns

Based on our current understanding, it is reasonable to conjecture that big challenges like curing cancer, reversing climate change, or colonizing Mars are more complex than we yet realize. Why is that reasonable? Because thinking about these "war efforts" is a lot like mulling over the anthropic principle: the idea that the universe seems to be arranged so as to be intelligible to humans.

One response to this observation centers around a "survivorship bias": a universe arranged in such a way as to allow for humans to evolve necessarily must be conducive to human inquiry. If the physical constants of the universe had been toggled any other way, we would not have come into existence, and thus would not be able to make conjectures about the universe in the first place. The point is that the universe was not designed *for us*; rather, we were naturally selected for it.

By the same token, there is little reason to believe in an anthropic principle of cancer, climate reversal, or Mars colonization. It's nice to think that an omniscient, omnipotent, and benevolent deity would not have left us with a challenge unless we could resolve it. But, if we're being honest with ourselves, we must accept that we inhabit a universe that has no need for us. The modern creed of progress tells us that with the right knowledge, all problems can be solved. But nature offers no assurances of when, let alone if, the "right" knowledge will be disclosed to us.

In a famous (though possibly apocryphal) anecdote, the American philosopher William James is confronted by a little old woman who tells him that, notwithstanding his fancy theories about heliocentrism, the Earth in fact rests on the back of a giant turtle. When challenged by James to explain what that turtle stands on, the women replies, "You're a very clever man, Mr. James, and that's a very good question. ... It's no use, Mr. James – it's turtles all the way down" (Ross 1967, p. iv).

In the case of cancer, each major advance in our knowledge has revealed something worse than just another turtle: ever-deeper complexity. Cancer cannot always simply be chopped out, irradiated, or poisoned away. It is deeply embedded in our physical nature as highly complex evolved beings. Our own immune systems are more powerful and sophisticated than anything we ourselves have ever created; and yet, cancer routinely evades or even hijacks this very system.

Indeed, the limitations in our understanding of the human immune system offers independent grounds to doubt expectations of a cure for cancer from recent advances in onco-technologies. As multicellular organisms, our bodies have a vast armory with which to respond to cancer, yet humans regularly die of the disease because it contrives to elude all of these countermeasures, adapting in a much shorter timeframe than the hundreds of millions of years that it took our immune systems to evolve. In cancer researchers' memoirs, one often glimpses hints of admiration for an adversary whose pluck enables it to survive and flourish in the determinedly hostile environment of the human body.

As Leslie Orgel, one of the foundational theorists of the origins of life, once quipped, "evolution is cleverer than you are." This observation, the second of "Orgel's Rules of Evolution," warns against the hubris of those who hype digital improvements to human hearts and eyes (Dunitz and Joyce 2013, p. 11). The sheer complexity of these evolved systems remains far beyond the scope of anything humans can design themselves.

Orgel's second rule also challenges the confidence of those who expect onco-immunology to cure cancer anytime soon. Although this therapeutic approach offers effective patches to help our immune systems detect and destroy cancer, we shouldn't be surprised to learn that tumors often adapt to these changes (Palucka and Coussens 2016, p. 1233; Graeber 2018). The question, then, is whether we can come to know enough about the immune system's anti-cancer measures to be able to outwit cancer's own rapid evolution within the body. When might we be entitled to reject Orgel's second rule and proclaim ourselves the cleverer ones?

One milestone, at least when it comes to our immune systems, would be to defeat all autoimmune diseases, a category that includes diabetes, multiple sclerosis, and lupus. These conditions all seem to result from misdirected attacks by our immune systems on our own healthy bodily tissues, but there is much that remains mysterious about them. The persistence of such diseases is a testament to our incomplete understanding of immunity.

Imagine a future society that cures Type 1 diabetes with a simple precision fix that switches off the misdirected attack on pancreatic islet cells, leaving the remainder of the immune system unaffected. Then society's onco-immunologists could credibly declare themselves cleverer than both our evolved immune systems and cancer's own evolved responses to it. Our present society, however, can make no such claim. The odd misguided act of vandalism notwithstanding, the Flame of Hope will continue to burn.

Our long struggle against cancer provides strong inductive grounds for believing that each new breakthrough will reveal yet another turtle. What we see as cutting-edge innovations, future observers may well regard as mere slings and arrows against an impregnable fortress wall.

Progress is always relative. If human biology can be converted into discrete information, that will be an impressive feat, and Kurzweil's vaticination will have been borne out. But it doesn't follow that cataloging such data will lead to a complete understanding of cancer, a cure for diabetes, or even a swift victory against the next pandemic disease.

After all, we have more socioeconomic data than any other society in history, yet economists still consistently fail to foresee looming meltdowns, nor can they even agree on the answers to basic, seemingly empirical questions, such as the effect of a particular tax policy on GDP growth.

The reification of data is, arguably, the defining feature of the information age. Entire industries rest on the assumption that more data—and more data-analytic power—will yield ever-more fruitful results. But the flaws in such thinking have long been apparent. As the cultural critic Neil Postman argued 30 years ago, the "accumulation of reliable information about nature," society, and the "human soul" has not automatically ushered in a better world (Postman 1992, p. 60). Just as there is the "problem of information scarcity," so too are there "dangers of information glut"—a problem that should be familiar to anyone who has tried to make sense of the world through Facebook or Twitter.

But even when we have wrestled information into control (through digitalization), even when we have squeezed it for insights (with AI), it is a mistake to conflate the resulting information with knowledge, let alone understanding. As Postman warned, the exponential growth of information can lead us down rabbit holes without actually telling us what we need to know. "Is it lack of information about how to grow food that keeps millions at starvation levels?," he asked. "Is it lack of information that brings soaring crime rates and physical decay to our cities? Is it lack of information that leads to high divorce rates and keeps the beds of mental institutions filled to overflowing?" (Postman 1992, p. 60).

The fact is that we know much less about the future than we think we do, and that will always be the case. We do not know what we don't know, and nor should we inflate the importance of our current understanding. The idea that we have reached a "critical mass" of knowledge and technological capacity inevitably will lead us to expect more than we should from rapid progress.

Forecasts based on the perception of exponential growth tend to lack a key piece of information: no one yet knows how difficult it will be to fight the next pandemic, cure cancer, reverse climate change, or colonize Mars. If anything, our progress against these problems may slow down, rather than accelerate.

High-Hanging Fruit

The tyranny of "times two"—and abstract ideas about exponential growth and improvement more generally—has long inspired overconfidence in the power of technology to solve difficult problems. But even if one accepts that information and communications technologies will continue to advance at a very rapid rate, one also must acknowledge the other side of the scale. There are factors weighing heavily against sustained progress, some of which are external, some of which may be endemic to the innovation process itself, and some of which concern our own awareness and expectations of progress. In forming rational expectations about future progress against big problems, one must account for the *net* effect of both "tailwinds" and "headwinds."

Consider the following video/board game analogy. Players of traditional role-playing games (RPGs) understand that their characters' powers increase exponentially as they advance through the game. But they also understand that these improvements are offset by improvements in the adversaries the game pits against them. It wouldn't be much fun to be obliterating Level 1 skeletons with your Level 100 Sword of Doom.

In games as in real life, the standard expectation is that the further one advances in a given field, the harder the challenges will become. In the context of a career, a quest, or any other long-term endeavor, we naturally presume that progress will lead to increasing difficulties. Why wouldn't the same intuition apply to science, technology, and medicine?

Though it seems counterintuitive in an age of techno-hype, it is not unreasonable to think that the march of technological and scientific progress will decelerate, even reaching something like a steady-state endpoint in certain domains. In 1996, John Horgan of *Scientific American* published *The End of Science*, arguing that humanity had already answered most of the "big questions" that it is capable of answering (Horgan 2015a). From the germ and big bang theories to evolution, relativity, and heliocentrism, fundamental discoveries of what he calls "pure science" can be made only once. Moreover, Horgan pointed out that "science itself, as it advances, keeps imposing limits on its own power:

> Einstein's theory of special relativity prohibits the transmission of matter or even information at speeds faster than that of light; quantum mechanics dictates that our knowledge of the micro-realm will always be uncertain; chaos theory confirms that even without quantum indeterminacy many phenomena would be impossible to predict; Kurt Gödel's incompleteness theorem denies us the possibility of constructing a complete, consistent mathematical description of reality. And evolutionary biology keeps reminding us that we are animals, designed by natural selection not for discovering deep truths of nature, but for breeding. (Horgan 2015a, p. xxx)

Horgan's point was not that "applied science"—the use of existing knowledge to solve problems—will ever come to an end. But when it comes to understanding "the universe and our place in it," we should expect to have figured most things out at some point. Not surprisingly, Horgan's book was controversial among professional scientists. "He's a very nice guy, and he wrote a very bad book," remarked the biologist Lynn Margulis. Similarly, the renowned biologist Stephan Jay Gould politely described Horgan's conclusion as "ridiculous," akin to declaring the "end of literature" (Horgan 2015b).

But in a 2015 reissue of the book, Horgan had a chance to look back on his initial forecast, and he found that it had largely stood the test of time. Not only had there been no major "revelations or revolutions" on the level of evolution or quantum mechanics, but some fields appeared to have regressed. Beyond string theory, which had ventured ever further out onto a metaphysical limb, the Human Genome Project had failed to deliver on its lofty promises to cure diseases like cancer.

Though many scientists bristle at Horgan's position, others are of a similar mind. The renowned British astronomer Martin Rees, for example, warns us against the anthropic fallacy, explaining that "any universe complicated enough to have allowed our emergence is for just that reason too complicated for our minds to understand" (Rees 2018). He also offers a word of caution about allowing our past achievements to instill hubris in the present. Though we have come to understand our place in the cosmos, "our grasp of some everyday matters that interest us all – diet for instance – is so meagre that expert advice changes each year. … it's complexity, not size, that makes phenomena baffling – the smallest insect is structured more intricately than a galaxy, and offers deeper mysteries" (Rees 2020).

Horgan and Rees are in good company. None other than Francis Bacon expected that pure science ("interrogatories about the facts of nature") would eventually have to end. In fact, he anticipated (rather wildly) that with a concerted effort and a commitment to strict empiricism, "it would take only a few years to discover all the causes and all science" (Bacon 1620, p. 88).

Moore's Nemesis

Similar headwinds have been discerned in key areas of scientific research and technological innovation. In the case of pharmaceutical R&D, it even has a clever name: Eroom's Law (the reverse of Moore's Law). Coined by industry analyst Jack W. Scannell and his colleagues in a March 2012 *Nature Review* article, Eroom's Law describes the fact that "the number of new drugs approved per billion US dollars spent on R&D has halved roughly every 9 years since 1950, falling around 80-fold in inflation-adjusted terms" (Scannell et al. 2012, p. 191).

Remember, Moore's Law suggests that the microchips of the machines we use to process medical data could rapidly become more powerful. But chips are not the only things that matter. In the role-playing game example, you may have the Sword of Doom but you won't long survive if you haven't diligently upgraded your armor, too.

Eroom's Law is one of many influences that bear on the problems like cancer. It suggests that we get progressively less therapeutic bang for each buck spent fighting the disease. In the example above—the halving of new drugs per billion dollars spent—the relationship between expenditure/investment and output is geometric, indicating that there could be an upper bound.

For example, here is a geometric series that results from dividing each increment by two, wherein the sum never exceeds one: $\frac{1}{2} + \frac{1}{4} + 1/8 + 1/16\ldots$. The implication is that for progress against disease to be maintained or accelerated, we will need to attend to the supply-side of dollars spent on research. But for the funding to keep up, researchers would have to convince investors to pour more and more money in for ever-smaller returns.

Scannell and his co-authors point to a number of possible causes for diminishing returns in drug research. But they start by emphasizing a distinction that is glossed over by those who would treat the human body as mere information to be catalogued and data mined. Unlike the relatively simple solid-state physics underlying integrated circuits, the biological subjects involved in drug research are deeply complex, dynamic systems for which our current understanding is still limited.

More narrowly, the authors identify a cousin to the low-hanging fruit phenomenon, which they deem the "better than the Beatles" problem. "Imagine how hard it would be to achieve commercial success with new pop songs if any new song had to be better than the Beatles, if the entire Beatles catalogue was available for free, and if people did not get bored with old Beatles records," they write. In the case of pharmaceuticals, where once-groundbreaking treatments eventually become generics, discovering even more effective drugs for even harder-to-treat conditions implies a steeper and steeper summit.

Another headwind is the "cautious regulator" problem, whereby the drug-approval process becomes more difficult over time. Whereas more red tape is added after every mishap, there are rarely occasions in which rules or standards are stripped back, resulting in a "tightening ... regulatory ratchet." When there are already a number of drugs on the market for a certain condition, regulators tend to set a higher bar for newer treatments geared toward the same purpose.

In an April 2020 follow-up to the original Eroom's Law paper, Scannell, along with three other co-authors, found that new drug approvals had risen sharply in the ensuing years, effectively reversing the trend toward diminishing returns (Ringel et al. 2020, p. 833). However, they attribute the increase to improved genetic information, which in turn has driven an uptick in drugs targeting rare, previously neglected diseases—for which the regulatory threshold was likely lower.

Ultimately, the authors conclude that these one-time factors do not refute the original hypothesis. The same "underlying forces that caused the prior decline, such as the 'betters than the Beatles' problem," remain, and will "again take hold once the effects of the recent improvements in understanding of disease biology and decision making wane."

Another potential factor in the reversal of Eroom's Law is a disconnect between approvals and actual health outcomes. In terms of cancer drugs, few new treatments have been able to match the remarkable success of imatinib (Gleevec), which was approved by the FDA in 2001 for treating chronic myeloid leukemia (CML). For most CML patients, imatinib is so effective as to represent something close to a cure, affording them as many additional quality-adjusted life years (QALYs) as they would have had without the disease (Pray 2008). Quite unexpectedly, CML was revealed as low-hanging fruit on the cancer tree. It yielded to our therapeutic efforts when many other cancers that might have seemed easier to treat at the time did not.

By contrast to imatinib, many other cancer drugs since then have gained approval after demonstrating only minimal median survival benefits over existing treatments. In calculating Eroom's Law, weighting cancer drugs according to the median additional quality-adjusted life years provided would likely yield a more disappointing picture.

In any case, although the "better than the Beatles" problem is somewhat unique to pharmaceuticals (owing to the structure of the intellectual-property regime and other factors), the broader economics literature on rates of technological innovation reveal similar findings. As in pharmaceuticals, R&D more generally has for decades been delivering less and less bang for the buck. While the number of researchers in the United States has grown significantly since the 1950s, innovative output has lagged. (The rate of innovative output can be expressed in various ways, including as the rate of patenting—patents per researcher—firm performance, or as total factor productivity (TFP) in the economy. Since not all innovation is patentable, and since patent laws have changed over time, the first metric is not necessarily dispositive.)

More to the point, it is not just that innovation is slowing down, but that it seems to be getting more difficult with the accumulation of ever more discoveries. One explanation for this is the "fishing-out" problem, which is another cousin to the problem of low-hanging fruit. In the 1990s, economist Samuel S. Kortum noted that as the stock of knowledge expands, so does the "technological frontier," which means that one must go ever further to find anything new under the sun (Kortum 1997, p. 1389). Hence, despite steady growth in the number of researchers, the rate of patenting remained roughly constant from the 1950s to the 1990s, and the total factor productivity (TFP) growth rate since the 1970s had declined.

Soon thereafter, the economist Benjamin F. Jones offered a slightly different interpretation, arguing that the problem is not just an expanding frontier but an increasing educational burden for individual researchers. To build on the break-throughs of past generations, one must learn everything they learned and then more. As a result, the age at which innovators make discoveries has been trending up, as has the size of research teams and the degree of specialization between individual researchers (Jones 2009).

Jones concludes that, "If a rising burden of knowledge is an inevitable by-product of technological progress," then "pessimistic predictions for long-run growth" are in order. While some "future revolution in science" could of course bring about a "fall in the burden of knowledge," the twentieth century's trends of "rapidly increasing R&D effort but flat TFP growth" do not inspire optimism (Jones 2009, p. 311).

Ironically, though not surprisingly, it turns out that Eroom's Law even applies to Moore's Law itself. In an April 2020 *American Economic Review* study, Nicholas Bloom, Charles I. Jones, John van Reenen, and Michael Webb note that the number of researchers needed to double the number of transistors on a microchip is now 18 times larger than it was in the early 1970s. Reaffirming earlier findings in the litera-ture, they show that while research effort has risen 23-fold since the 1930s, research productivity has "fallen by an even larger amount, by a factor of 41" (Bloom et al. 2020, p. 1105).

The authors also look specifically at cancer research and its effects on cancer outcomes in the United States. They find that the years of life saved per 1000 people (based on current life expectancy at birth and at age 65) in proportion to annual research efforts peaked around 1990, and then declined by 40% in the 2000s. All told, they report that "between 1975 and 2006, research productivity for all cancers declines by a factor of 1.2 using all publications and a factor of 4.8 using clinical trials" (Bloom et al. 2020, p. 1126).

In practice, this decline means that between 1985 and 2006, "the number of years of life saved per 100,000 people in the population by each publication of a clinical trial related to cancer *declined from more than 8 years to just over one year*" [Emphasis ours]. And the diminishing returns in the case of breast cancer were even worse—from 16 years per trial in the 1980s to less than 1 year as of 2006 (Bloom et al. 2020, p. 1127).

In addition to semiconductors and health care, Bloom and his co-authors find diminishing returns from research in agriculture, cross-sector firm performance, and the aggregate economy, concluding that "research productivity is declining at a substantial rate in virtually every place we look" (Bloom et al. 2020, p. 1133). Again, this phenomenon should not come as a surprise. One should expect things to get harder over time, and similar long-running patterns have been discerned world-wide (Evenson 1984).

By the same token, one cannot just assume that economic and technological progress will continue apace. Techno-optimists, lauding the splendors of Moore's Law, tend to expect as much, yet it is a mistake to view the linear progression of microprocessors as a law (like gravity), when really it is just an observation—and, perhaps, a self-fulfilling prophecy.

As Bloom and his co-authors emphasize, maintaining the prophecy requires more and more researchers at an ever-greater cost. And yet, the number of available qualified researchers is a function of population, which is shrinking in most indus-trialized countries. The more pronounced this trend becomes, the greater the threat to future knowledge creation and innovation (Jones 2022).

Where the Rubber Meets the Road

The standard retort to the diminishing-returns argument is to point out that flat returns in one domain do not imply flat returns across the board. After all, when electric vehicles powered by lithium-ion batteries become the new universal norm, one wouldn't expect for there to still be continuous improvements in internal-combustion-engine technologies. There is not one "S-curve"—flat growth followed by exponential growth followed by a flattening of time—but many. The telegraph S-curve gives way to the telephone S-curve which gives way to the smart phone S-curve.

But it is important to note that this argument relies on inductive evidence from the history of technology more broadly. When it comes to more specific questions like the War on Cancer, the same inductive lens shows a cloudier picture. Historically, one "game-changer" after another has fallen short of expectations. From Sidney Farber's breakthrough with antifolates in the 1940s to the marginal gains from gene therapies today, there has yet to be a discovery in cancer research that drives out-comes on a scale comparable to technological revolutions like electrification in the twentieth century and digitalization in the twenty-first.

Again, the point is to think carefully about what our rational expectations should be with respect to the big problems identified by futurists. It is always possible that some discovery could change everything tomorrow. Yet when all the factors dis-cussed in the previous chapters are weighed in the balance, we would argue that one should not expect imminent, decisive breakthroughs.

A good way to clarify one's rational expectations is to think in terms of placing a bet. When someone predicts that humanity will develop "the cure" for cancer or colonize Mars, it is appropriate to ask him what he would bet on it. Making a bet requires that he identify a specific line of research and accept certain limiting condi-tions, not least an agreed upon target date for settling up. Specifying these terms and placing a real wager helps to expose the gap between nice words and intellectually responsible claims about the future.

Betting on the future of something like cancer may seem uncouth. But as a thought exercise, it is one of the best ways to clarify statements of prudential and moral rationality that are otherwise lying beneath the surface of our subconscious thought. We place bets in situations when our expectations of future results differ with those of others. When we put down money (or some other good), we stake something of value with the hopes of achieving future rewards. But the most we can say about these rewards is that we consider their attainment to be possible (one can, of course, wager on the impossible; but we wouldn't advise it).

As a collective, a society can also stake something of value on claims about the future. That is what Nixon did when he declared war on cancer. He ventured things of value—including money and the brain power of researchers—on a successful permanent result against cancer. Similar reasoning is at the center of debates about anthropogenic climate change. In recent years, the phenomenon of climate-science denialism has finally fallen out of favor. Few serious people in positions of power refute the basic fact human activities are causing the planet to warm and the climate to change. And yet, when we look past the words offered by politicians and policy-makers, we find that we have effectively placed a wager on climate-change denial.

Every politician can say that they are fully convinced that climate change poses a serious threat. But what have we done about it? By not instituting a carbon tax, investing in renewable energies, and so forth, we have implicitly placed a bet on the less threatening futures in the range of possibilities postulated by climate deniers. The mechanisms whereby greenhouse-gas emissions change the climate are well supported scientifically. But climate scientists hasten to remind us that, owing to the sheer complexity of the system that is being modeled, there can be little absolute certainty of the precise effects of climate change over time. The worst-case scenarios are truly catastrophic; but the standard models also offer tamer scenarios that would allow humanity to avoid making hard sacrifices in the present. Naturally, it is these potential outcomes that special interests and opportunists tend to seize on.

Science does not traffic in the certainties claimed by religion (or, at least, it isn't supposed to). For all we know, a climatological mechanism hitherto undescribed by science might be discovered, pointing the way to an easier solution to climate change. If that happened, the deniers could feel justifiably smug, and the rest of us would not have to beat ourselves up over the moral failure of doing nothing. But that's beside the point, because given what we do know, humanity's survival on the planet isn't something we should bet on.

References

Agar, Nicholas. 2015. *The Sceptical Optimist: Why Technology Isn't the Answer to Everything.* Oxford University Press.

Bacon, Francis. 1620/2000. *The New Organon.* Cambridge University Press.

Bloom, Nicholas, Charles I. Jones, John Van Reenen, and Michael Webb. 2020. Are Ideas Getting Harder to Find? *American Economic Review.*, April 110 (4): 1104–1144.

Brynjolfsson, Erik, and Andrew McAfee. 2014. *The Second Machine Age: Work, Progress, and Prosperity in a Time of Brilliant Machines*. W. W. Norton.

De Waal, Alex. 2021. *New Pandemics, Old Politics*. Polity.

DeVita, Vincent T. 2002. A Perspective on the War on Cancer. *The Cancer Journal*, September–October 8 (5): 354. https://doi.org/10.1097/00130404-200209000-00002.

DeVita, Vincent T., and Elizabeth DeVita-Raeburn. 2016. *The Death of Cancer: After Fifty Years on the Front Lines of Medicine, a Pioneering Oncologist Reveals Why the War on Cancer Is Winnable – and How We Can Get There*. Sarah Crichton Books.

Diamandis, Peter. 2017. 3 Dangerous Ideas from Ray Kurzweil. Singularity Hub, 10 November.

Dizikes, Peter. 2018. Thomas Friedman Examines Impact of Global "Accelerations.". *MIT News*. 2 October.

Dunitz, Jack D. and Joyce, Gerald F. 2013. *Biographical Memoirs: Leslie E. Orgel, 1927–2007*. National Academy of Sciences. http://www.nasonline.org/publications/biographical-memoirs/memoir-pdfs/orgel-leslie.pdf. Accessed 12 Oct 2020.

Evenson, Robert. 1984. International Invention: Implications for Technology Market Analysis. In *R&D, Patents and Productivity*, ed. Z. Griliches. University of Chicago Press.

Friedman, Thomas. 2016. *Thank You for Being Late; Finding a Job, Running a Country, and Keeping Your Head in an Age of Accelerations*. Farrar, Straus and Giroux.

Gordon, Robert J. 2016. *The Rise and Fall of American Growth: The U.S. Standard of Living Since the Civil War*. Princeton University Press.

Gottlieb, Scott. 2017. Remarks from FDA Commissioner Scott Gottlieb, M.D., as prepared for oral testimony before the U.S. Senate Committee on Health, Education, Labor & Pensions Hearing, "Implementation of the 21st Century Cures Act: Progress and the Path Forward for Medical Innovation.". *FDA News Release*. 7 December.

Graeber, Charles. 2018. *The Breakthrough: Immunotherapy and the Race to Cure Cancer*. Twelve.

Haseltine, William A. 2020. What If There's No COVID Vaccine? *Project Syndicate*. 24 July.

Horgan, John. 2015a. *The End of Science: Facing the Limits of Knowledge in the Twilight of the Scientific Age*. 2nd ed. Basic Books.

———. 2015b. Was I Wrong About "The End of Science"? *Scientific American*. 13 April.

Hosny, Ahmed, Chintan Parmar, John Quackenbush, Lawrence H. Schwartz, and Hugo J.W.L. Aerts. 2018. Artificial Intelligence in Radiology. *Nature Reviews Cancer* 18 (8): 500–510. https://doi.org/10.1038/s41568-018-0016-5.

Jones, Benjamin F. 2009. The Burden of Knowledge and the "Death of the Renaissance Man": Is Innovation Getting Harder? *The Review of Economic Studies*., January 76 (1).

Jones, Charles I. 2022. The End of Economic Growth? Unintended Consequences of a Declining Population. *American Economic Review* 112 (11): 3489–3527. https://doi.org/10.1257/aer.20201605.

Keynes, John Maynard. 1930/1963. Economic Possibilities for our Grandchildren. *Essays in Persuasion*. W. W. Norton & Company.

Kortum, Samuel S. 1997. Research, Patenting, and Technological Change. *Econometrica*, November, No. 6: 1389–1419.

Kurzweil, Ray. 1999. *The Age of Spiritual Machines*. Viking.

Le Fanu, James. 2011. *The Rise and Fall of Modern Medicine*. 2nd ed. Abacus.

Moore, Gordon E. 1965. Cramming More Components Onto Integrated Circuits. *Electronics*: 19. April 19.

Palucka, Karolina A., and Lisa M. Coussens. 2016. The Basis of Onco Immunology. *Cell*, 10 March 164 (6): 1233–1247.

Postman, Neil. 1992. *Technopoly: The Surrender of Culture to Technology*. Knopf.

Pray, Leslie A. 2008. Gleevec: The Breakthrough in Cancer Treatment. *Nature Education* 1 (1): 37.

Raza, Azra. 2019. *The First Cell: And the Human Costs of Pursuing Cancer to the Last*. Basic Books.

Rees, Martin. 2018. What Are the Limits of Human Understanding? *Prospect*. 13 November.

————. 2020. Why I'm Skeptical About the Unique Power of the "Scientific Method.". *Prospect*. 29 January.

Reese, Byron. 2013. *Infinite Progress: How the Internet and Technology Will End Ignorance, Disease, Poverty, Hunger, and War*. Greenleaf Book Group Press.

Ringel, Michael S., Jack W. Scannell, Mathias Baedeker, and Ulrik Schulze. 2020. Breaking Eroom's Law. *Nature Reviews Drug Discovery*. April 19: 833–834. https://doi.org/10.1038/d41573-020-00059-3.

Ross, John Robert. 1967. *Constraints on Variables in Syntax*. MIT dissertation, September. https://dspace.mit.edu/handle/1721.1/15166. Accessed 21 May 2021.

Scannell, Jack W., Alex Blanckley, Helen Boldon, and Brian Warrington. 2012. Diagnosing the Decline in Pharmaceutical R&D Efficiency. *Nature Reviews* 11, March: 191–200.

Topol, Eric. 2019. *Deep Medicine: How Artificial Intelligence Can Make Healthcare Human Again*. Basic Books.

Tucker, Toph. 2016. The Exponential Curve Has No Knee. Toph Tucker's Block., 21 November. https://bl.ocks.org/tophtucker/bc2d6937ed83c9e67fab592eef70e0a5. Accessed 12 Mar 2020.

Chapter 7
The Hand-Off

Contents

In 1904, the American historian Henry Adams (the grandson of John Quincy Adams and the great-grandson of John Adams) postulated his own kind of Moore's Law to describe the rapid progress of the previous centuries. World coal output, he noted, had roughly "doubled every ten years between 1840 and 1900, in the form of utilized power, for the ton of coal yielded three or four times as much power in 1900 as in 1840." He used the improved marginal efficiency of coal as a proxy for technological and scientific progress, which he regarded as a potentially inexhaustible force. But he foresaw problems:

> If it behaved like an explosive, it must rapidly recover equilibrium; if it behaved like a vegetable, it must reach its limits of growth; and even if it acted like the earlier creations of energy – the saurians and sharks – it must have nearly reached the limits of expansion. If science were to go on doubling or quadrupling its complexities every ten years, even mathematics would soon succumb. An average mind had succumbed already in 1850; it could no longer understand the problem in 1900. (Adams 1996, p.496)

Notwithstanding Adams's rhetorical flourishes (surely mathematics can indeed keep up with rising complexity), he had identified a problem that we touched on in the previous chapter. He went on to conclude that further accommodation of scientific progress would "require a new social mind." For thousands of years, the human mind had adapted to the growth of knowledge, "and nothing yet proved that it would fail to react." But at some point, he surmised, "it would need to jump" (Adams 1996, p. 498).

Adams ended his historical reverie there, leaving the reader to infer what he might have meant by a "jump." Assuming he wasn't harboring plans for a eugenicist project to improve humanity's cognitive abilities through breeding, it stands to reason that he was thinking about technological extensions of human power. After all, the Baconian tradition has always depended on technology to open up new epistemological avenues. Without the invention of the telescope and microscope, we never could have made any serious advances in astronomy/cosmology or biology.

Historically, technology has been the key variable determining what someone can know empirically at any point in time. Before scientists were able to see and confirm phenomena unperceivable to the raw human senses, the best they could offer was metaphysical speculation or flights of fancy like Francis Godwin's *The Man in the Moone* (1638).

Each generation of technology can take us only so far. Once scientists had discovered and categorized the basic building blocks of biological life and the cosmos, once they had organized the existing corpus of knowledge within mathematical models, they found that many additional questions remained. The deeper they probed, the more they realized how much more there was still to learn. Unlike the consistent, predictable mechanics of an apple falling from a tree, quantum and genetic phenomena are subject to a degree of randomness, indeterminacy, and uncertainty that demands a new epistemology.

Similarly, within complex systems (from the human body to a modern economy) understanding isolated phenomenon (a single cancer cell; a single viral particle; a single consumer) no longer will suffice. Such systems feature emergent properties that can be understood only at the level of the system itself. Or, perhaps they cannot be understood at all. As of the early twenty-first century, the mathematician Edward R. Dougherty contends:

> … it has become apparent that the epistemology that began with Galileo, took shape with Isaac Newton, and came to fruition in the first half of the Twentieth Century with Niels Bohr, Hans Reichenbach, and others cannot support the desire to model complex systems. Across disciplines, scientists and engineers want to gain knowledge of large-scale systems composed of thousands of variables interacting nonlinearly and stochastically, often over long time periods. This massive complexity makes the standard modes of discovery and validation impossible. (Dougherty 2016, p. 8)

The difficulties posed by complexity are especially consequential in medicine, which involves physiological systems comprising trillions of cells interacting with one another through innumerable pathways and across multiple parameters. Of all diseases, cancer is perhaps the most complex of all. As the late Kenneth L. Mossman explains, "Advanced cancer defies prediction and is so variable from patient to patient that treatment outcomes are difficult to forecast even in patients with the same diagnosis" (Mossman 2014, p. 141).

A similar problem attends pandemics. As Alex de Waal notes, each new pathogen—from cholera, influenza, and HIV to yellow fever, Ebola, and COVID-19—has launched a search for prevention and cure that takes "scientists into new terrain they could not have anticipated" (De Waal 2021, p. 5).

Moreover, at the nadir of the COVID-19 pandemic, we found our predictive capacities challenged by social and political complexities that reached beyond the basic biochemistry of SARS-CoV-2. Different communities and countries responded to the challenge in a variety of unpredicted—and possibly unpredictable—ways.

Much of our previous "knowledge" about pandemics had come from movies like Steven Soderbergh's 1997 film, *Contagion*, where a lethal virus arrives, kills a lot of people, overwhelms hospitals, brings civil unrest, and then is swiftly defeated following the discovery of a vaccine. In the movie, things largely return to how they were before, and the imagined pathogen, MEV-1, is objectively worse than SARS-CoV-2, with its terrifying R0 (a measure of contagiousness) of 4.0 and a mortality rate of 25–30%.

More to the point, the *Contagion* virus is so lethal and so contagious that it sends a straightforward message to everyone. Unlike in the COVID-19 pandemic, almost nobody in the movie dismisses the virus as a "little flu" that their immune systems can easily overcome. By contrast, SARS-CoV-2 has a Jekyll and Hyde dynamic, exhibiting no symptoms in many people but acting, initially, as a mass killer in retirement communities. This ambiguity, combined with the complications of actually immunizing the entire world, suggest that even with very effective vaccines, the threat to some populations will linger on.

Adversaries with Evolution on Their Side

One reason why cancer and COVID-19 pose such a problem is that each is empowered by evolution. A chemotherapy that kills 99% of a patient's cancer cells predictably leaves behind a tumor constituted of cells that are resistant to that therapy. Similarly, SARS-CoV-2 "responds" to the rollout of highly effective vaccines by producing highly "effective" variants of concern that can spread more easily than their precursors.

From late 2019 to late 2021, much of the Greek alphabet had already been used to name COVID-19 variants of interest. In late-2021, the Omicron variant was ravaging many countries around the world, and it still reigned three years later. There were many new subvariants of Omicron, of course, but none sufficiently different to be dubbed Pi by the World Health Organization. Despite this deceleration, we might eventually arrive at omega, whereupon we will have to look beyond the limits of the Greek alphabet to extend the taxonomy. As the renowned HIV researcher William A. Haseltine observed in June 2021:

> Looking across the topology of the SARS-CoV-2 genome, there appear to be endless possibilities for the generation of novel viable variants. Like the influenza viruses that never seem to run out of ways to evade the previous year's vaccines, cold-causing coronaviruses evade our immune responses every year. (Haseltine 2021)

The evolutionary process that generates therapy-resistant tumor cells and, potentially, vaccine-resistant SARS-CoV-2 variants might appear slow compared to the exponentially progressing medical technologies that have been brought to bear on these threats. But, as we explained in Chap. 6, it is a mistake to assume that exponential technological progress will promptly solve all our problems. The celebrated "bends in the knee" exist wholly in the eyes of the beholder.

To be sure, our exponential technologies would appear to be advancing much faster than any virus or tumor can through the plodding process of evolution. But as Ray Kurzweil himself reminds us, evolution is also an exponential process, because the beneficial mutations that are preserved by natural selection build on each other. And this is not just theoretical. We have seen first-hand, in 2021, how the Delta variant of the coronavirus quickly came to be the dominant strain globally, overcoming public-health measures that had worked just fine for earlier variants. Later the same year, the Omicron variant began taking over, moving the goal posts again. That process unfolded neither predictably nor incrementally.

Alien Intelligence

The delta wave of the coronavirus points to what can happen when we oversimplify how we think about exponential technological progress. Looking ahead, we should bear this danger in mind when considering solutions to big problems based on artificial intelligence—which is by far the most celebrated manifestation of exponential technological progress.

First, we should clarify what we mean by AI. As Alan Turing noted in his famous 1950 paper, "Computing Machinery and Intelligence," the question is not whether machines can think; it is whether we as a culture can "speak of machines thinking without expecting to be contradicted." For his part, Turing expected that we would do so by "the end of the century" (Turing 1950, p. 442).

There are two ways we can think about intelligent machines. The first pertains to machines capable of exhibiting stereotypically human intelligent behavior. It is this facet of AI that the imitation game in Turing's famous test was designed to pick up. Hence, Data, the fictional android in *Star Trek: The Next Generation*, can easily pass the Turing Test.

Inspired by Turing, researchers have set a goal of achieving Artificial General Intelligence (AGI)—the capacity to understand or learn any intellectual task that a human being can perform. But this quest has gone through many periods of hype followed by "AI winters," owing to researchers succumbing to the horizon bias. (To their credit, computer scientists seem not to forget these failed forecasts, whereas some others are all too happy to forget every past failed forecast of a cure for cancer.)

But while AGI continues to elude us, a second way of thinking about AI has unmistakably taken hold. We experience AI—if not AGI—every time we are targeted with uncannily accurate advertisements. The algorithms that deliver this content operate through machine learning and have been getting better and better at finding lucrative patterns in our online behavior. We can predict that they will continue to improve. There are no AI winters in the development of machines that are increasingly good at selling us stuff.

Moreover, with this function satisfied, there has been less of a need for AIs that can pass the Turing Test. Facebook's machine learners have no capacity to maintain indistinguishably human-like conversations with us because they don't need it. The company's stockholders are content with a superhuman capacity to surveille us and target us with ads.

Here, we can see the tension between the two conceptions of AI. AGIs are supposed to be intelligent in a way familiar to every human being. But today's machine learners are intelligent—indeed superintelligent—in distinctively inhuman ways. Not even the brightest minds at Google can explain precisely why your combination of internet searches elicited that particular advertisement for a Caribbean cruise (as opposed to the innumerable alternate ads that could have been selected).

When machine-learning algorithms spit out a particular finding, it tends to come without "explainability." An AI might detect a correlation between consuming radishes and rates of cancer within some subset of the population, but it cannot tell us how or why it came to that conclusion. Moreover, AI developers have yet to come up with a solution to this "black box" problem, because the lack of explainability is a structural feature of how deep learning works.

Hence, following DeepMind's announcement that its AlphaFold machine-learning tool had predicted the 3D structures of 98% of human proteins, an editorial in *Nature* cautioned that "Although AI in science and technology is good at producing accurate results, it doesn't (at least for now) explain how, or why, those results happened." And so, "there is still work to be done to unlock the science – the essential biology, chemistry and physics – of how and why proteins fold" (Editorial 2021).

Gamblers who decide that they've divined the pattern in slot machine pay-outs tend to end up broke. There doubtless are such patterns, but they would not be obvious or even discernible to humans. Similarly, we can imagine an AI capable of spotting the onset of a pandemic or a climate-driven event—such as the unprecedented heat wave that struck Canada and the American northwest in mid-2021. But, given the number of variables in play, its inferences probably will not follow pre-established rules that are easily discernible to us.

Paging Dr. Robot

Still, if AI could deliver victory in the War on Cancer, a fix for climate change, or a way to prevent future pandemics, should we really care how it carries out those tasks? Google and Facebook have become some of the most valuable companies in history by handing over the ad business to black-box algorithms. What would it mean to respond to civilizational problems by handing them off to machines, too?

Until recently, the historian Yuval Noah Harari notes, "humans were special and important because … they were the most sophisticated data processing system in the universe, but this is no longer the case." Just think of the experience of using a navigation app like Google Maps or Waze. "On the one hand," Hariri explains, these "amplify human ability – you are able to reach your destination faster and more easily. But at the same time, you are shifting the authority to the algorithm and losing your ability to find your own way" (Solon 2017).

The same, of course, could be true of our quest to tackle the Big Cs. It is easy to imagine a time, not too long in the future, when it will no longer be a human's job to deal with pandemics or cancer; indeed, in such a scenario, we mere mortals would be actively discouraged from acting as extra "cooks in the kitchen."

We know that cancer will never be eradicated. But what if it was no longer therapeutically useful for us ever to hear the word, other than as a constellation of stars? In a techno-optimistic vision, people's bodies would still develop cancers, but the machines of a fully mature digital age—what we'll refer to broadly as Dr. Robot—would nip it in the bud without our even knowing it.

In this future, cancer could continue to work as a metaphor, but we would no longer need metaphors for it. We would not have "cured" it, exactly, but the war would be over. Cancer, as an idea, would likely haunt our historical memory the way the Black Death does today. For example, the cancer metaphor might be hurled at social trends that deserve disapprobation, such as a wave of students cheating on exams or a loss of confidence in public institutions. One could speak of a "cancer in higher education" without having to worry about offending anyone who is actually suffering from the disease.

After enough time has passed in this possible future, cancer and pandemic threats might fade entirely from our collective consciousness, such that we needn't worry our pretty little heads about them. We won't even have to think about whether our car is climate-friendly, because these will be the only ones that machine learners build or allow us to acquire.

What would this mean for the human condition? Those who entertain notions of such a future rarely consider the full implications of handing over so much to machines. If human cancer specialists are no longer the ones in charge, this major field of medicine will essentially have passed into a post-human phase.

Yet in thinking about the moment of transition from human to machine oversight of cancer, there are two important points to consider. First, as we've seen, the horizon bias tends to make cures for cancer seem more imminent to the best informed than to the lay public. People would do better to rely on historical trends to make inductions about the progress against the disease. As such, we should anticipate that any transformation in our species' engagement with cancer would itself be subject to the same effect, at least for a while. We cannot say with any precision when the hand-off of our civilizational challenges to machines will occur. But we can say, if current trends persist, that it is coming.

Moreover, for Dr. Robot to become the new normal, it need not demonstrate omniscience; it just has to prove more effective than human physicians. Given humanity's epistemological limitations and the current trajectory of technological innovation, one could say with some confidence that the transition to Dr. Robot is coming at some point, even if one cannot be too sure of when. Similar projections can be made when thinking about climate change. We have more than enough scientific evidence to know that there is a problem, but we don't know precisely when any given coastal city will be subsumed by rising sea levels.

When John Maynard Keynes famously quipped that, "In the long run we are all dead," he was cautioning against focusing too much on the distant future, lest one overlook the potential for significant booms and busts along the way. But one can flip this reasoning on its head: we mustn't lose sight of the general trajectory by obsessing over the specifics or cyclical movements of the here and now.

Again, this is not a prediction. Often, books about science and technology conclude with speculative misadventures. If you are an expert in space propulsion, it is all too tempting to extrapolate from your technical work and explore the possibility of flights to Mars or even to Alpha Centauri. The same pattern is common in books about cancer. Once an author has gone through the serious science of systems biology and immunotherapy, she is free to speculate about a world in which nanobots linked to a central neural network identify and remove cancer cells as soon as they appear.

We are not oracles. But we believe it is important to think through this *possible* AI-centered future that one often hears described. When a doctor warns a smoker about the risks of lung cancer, she is not making a forecast about that patient's future. In fact, statistically, it is likely that the smoker won't get lung cancer (about 10–20% of long-term smokers get lung cancer; Burns 2000; Crispo et al. 2004). But the doctor's purpose is to make the patient acknowledge that the lung-cancer future is a credible possibility.

By the same token, we contend that a scenario in which cancer and other major problems have been handed off to machines is not only possible but increasingly likely if current trends continue.

Not Like Riding a Bike

In August 2013, *The Economist* repeated a familiar tale about the future of automation in commercial aviation: "One day in the not too distant future, so the hoary old story goes, airliners will have only two crew members on the flight deck – a pilot and a dog. The pilot's job will be to feed the dog. The dog's job will be to bite the pilot if he touches the controls" (The Economist 2013, no page).

The use of "hoary" suggests a dismissive attitude, implying that the magazine does not consider this possible future to be close enough to worry about. "Despite all the talk about drone-like autonomous passenger planes," the author declares, "cockpit automation is nowhere near capable enough to manage without human pilots on the flight deck. It is doubtful whether the technology – at least, as it is currently configured – will ever let that happen."

This deliberate myopia about the impact of automation on work is mistaken. Those who think seriously about the future of automation must allow for a future in which humans are barred from the cockpit. And those who think seriously about the future of the climate should consider a possible future in which humans are removed from the equation. In this future, human experts would not only have nothing to add but their mere presence in the consultation room would be problematic.

In the case of aviation—and, more recently, autonomous vehicles—one by-product of the machine taking over is that the humans become deskilled for lack of practice. In these instances, automation essentially becomes a self-fulfilling prophecy. But in the case of Dr. Robot, the issue isn't just de-skilling; it is about the application of knowledge. The human oncologists of 2100 would doubtless know far more about treating cancer than the best human oncologists of the 2020s. But for cancer patients in 2100, the choice will not be between listening to the best-informed oncologists of their time or the best-informed oncologists of 80 years earlier. It will be between placing their fates solely in the hands of machines and granting human doctors some say in the matter.

The question for this future scenario is whether even very knowledgeable humans would be able to add anything useful. Enlightened amateurism might be celebrated, but only because it would be kept out of life-or-death matters. Consider a non-physician who goes and masters every textbook and Wikipedia page that exists on coronary bypass surgery. It is safe to assume that no patient on the operating table will want this individual standing by to offer insights. All of the information she has studied will doubtless have made her very knowledgeable in objective terms, but one hopes that the actual surgeons still would know more. Her insights would serve as temporary distractions.

By recognizing the authority of a fully wrought Dr. Robot, patients and doctors alike would effectively be placing their faith in the black box of the algorithm, accepting as true those insights that they can neither see nor even interrogate. In this possible future, a human doctor worrying his pretty little head about the inscrutable complexities of the disease could only serve as a source of unnecessary confusion.

To be sure, many commentators argue that there will be more of a brokered compromise between humans and machines, with the latter augmenting rather than replacing the former. Erik Brynjolfsson and Andrew McAfee, for example, tell us that, "If the world's best diagnostician in most specialties – radiology, pathology, oncology, and so on – is not already digital, it soon will be." But physicians needn't worry, because, "Most patients … don't want to get their diagnosis from a machine" (Brynjolfsson and McAfee 2017, p. 124). There are obvious rebuttals to this argument, not least the fact that most patients in modern welfare states aren't the one's paying the bill, and thus wouldn't be the ones deciding ultimately who delivers their diagnosis. If the machine option is radically less expensive at scale, that is what governments and insurers will opt for.

Still, the idea of a human-machine power-sharing arrangement has also been endorsed by one of the early casualties of artificial intelligence: Garry Kasparov, the world chess champion whom IBM's Deep Blue defeated in 1997. In a February 2020 interview with *Wired,* Kasparov averred that it is "the role of a human" to figure out what any given machine "will need to do its best. At the end of the day it's about combination" (Knight 2020, no page).

> For instance, look at radiology. If you have a powerful AI system, I'd rather have an expe-
> rienced nurse than a top-notch professor [use it]. A person with decent knowledge will
> understand that he or she must add only a little bit. But a big star in medicine will like to
> challenge the machines, and that destroys the communication. (Knight 2020, no page)

The implication of this view is something of a paradox: humans can remain in charge as long as they show some humility before the machine. Kasparov adds:

> I describe the human role as being shepherds. You just have to nudge the flock of intelligent
> algorithms. Just basically push them in one direction or another, and they will do the rest of
> the job. You put the right machine in the right space to do the right task. (Knight 2020,
> no page)

The machine-learning researcher Pedro Domingos also offers reassurances for the future. Even though he expects AI to assume responsibility for many cognitive tasks, he assures us that humans will control the "last mile." It will still be up to us to pick between various options, even if the choices on offer have been selected by algorithms" (Domingos 2015, p. 12). In effect, humans will be like CEOs, giving a thumbs up or down to the proposals placed in front of us.

The Last Mile Is the Longest

These are the optimistic scenarios, but a more realistic outlook would be much darker. Because of the black-box problem, ceding agency to deep neural networks is not just another instance of human tool use. As the mathematician Steven Strogatz observes, human chess players often end up frustrated and confounded when they try to improve their games by learning from Google DeepMind's AlphaZero program. Since "the algorithms can't articulate what they're thinking," he writes, "We don't know why they work, so we don't know if they can be trusted. AlphaZero gives every appearance of having discovered some important principles about chess, but it can't share that understanding with us" (Strogatz 2018).

This problem is partly a consequence of the manner by which AlphaZero "learned" chess in the first place. Today's machine learners beat the best human chess and Go players by practicing an enhanced version of what Bobby Fischer did to become the world's best chess player. The young Fischer spent his days playing chess games against himself. AlphaGo beat the best human Go players by doing a radically enhanced version of that. The difference is that, if Fischer had wanted to, he could have described his memories of what was going through his head when he defeated the Soviet champion Boris Spassky. There is little chance of witnessing such introspective awareness from AlphaGo.

In the case of Deep Blue, IBM tested the algorithm against very good human players, and the programmers then responded with tweaks to fix defects in the machine's play. As a result, much of Deep Blue's strategic decision-making was comprehensible to humans. By contrast, AlphaZero mastered the game through self-play and "reinforcement learning," creating a log of examples through which to evaluate future choices. Following from a set of basic rules written by human programmers, it was able to play itself over and over—as many as 44 million times in the space of just 9 h (Silver et al. 2018).

Broadly speaking, this approach is the same as the one used to "train" autonomous vehicles (AVs). Humans guide the car through its initial lessons on the road, and the data from millions of "captcha" sessions (when one is asked by some website to identify the photos with crosswalks in order prove that one is human) are fed to the learning algorithm. But beyond a certain point, the car is increasingly teaching itself from real-world experiences. It will start to identify correlations. For example, to borrow a scenario from the AI researcher Stuart Russell, if the cars in the left-turn lane are stopped even though they have a green turn arrow, something might be amiss in the intersection ahead; the wise AV traveling unhindered in the center or right lane will slow down, even if there are no discernible obstacles in its immediate line of sight (Russell 2019).

One consequence of this self-learning method is that the machine very quickly "leaves the nest," so to speak: Its decisions are no longer intelligible to human observers. When Kasparov studied AlphaZero's moves, he reported seeing echoes of his own aggressive style of gameplay. But such claims cannot even be tested. Unlike with Deep Blue, where the programmers may indeed have introduced Kasparov-like instructions, AlphaZero has not had any meaningful contact with Kasparov's chess mind. To find something familiar in its gameplay is a bit like spotting the shape of an elephant in a cloud floating overhead.

Dr. Robot would exhibit the same epistemic distance as these other decision-making algorithms. It might well provide us with superior responses to pandemics. But we won't really know how it arrives at its suggestions. In some cases, the algorithm's output will be familiar to human experts. Knowledgeable humans will nod approvingly. But soon enough, Dr. Robot, too, would leave the nest, never to return. It would offer suggestions that never would have occurred to even the most knowledgeable human expert.

Imagining a future in which "AlphaZero has evolved into a more general problem-solving algorithm," Strogatz concludes:

> For human mathematicians and scientists, this day would mark the dawn of a new era of insight. But it may not last. As machines become ever faster, and humans stay put with their neurons running at sluggish millisecond time scales, another day will follow when we can no longer keep up. The dawn of human insight may quickly turn to dusk.
>
> Suppose that deeper patterns exist to be discovered – in the ways genes are regulated or cancer progresses; in the orchestration of the immune system; in the dance of subatomic particles. And suppose that these patterns can be predicted, but only by an intelligence far superior to ours. If AlphaInfinity could identify and understand them, it would seem to us like an oracle.
>
> We would sit at its feet and listen intently. We would not understand why the oracle was always right, but we could check its calculations and predictions against experiments and observations, and confirm its revelations. Science, that signal human endeavor, would reduce our role to that of spectators, gaping in wonder and confusion. (Strogatz 2018)

But why must this be the case? Why can't we just preserve a place for humans by designing Dr. Robot to explain its decisions to us? After all, famous popularizers of the hard sciences have clearly mastered this skill. In his best-selling *A Brief History of Time*, Stephen Hawking presented a wealth of highly technical knowledge in language and terms easily understood by the lay reader. If Hawking can do it, why can't a machine? The algorithm would just need to supplement its vast troves of data about cancer with knowledge about human comprehension. With a bit of extra programming, it could "learn" to explain its decisions to us.

It's a nice thought, but we shouldn't hold out hope for this kind of collaborative relationship. As we've seen, a superintelligent medical algorithm would likely remove the physician from the equation, even as a mediator, which means that its explanations would need to make sense not to a knowledgeable oncologist but to the average lay patient. But more to the point, a machine learner that tries to convey its discoveries to human beings would face an obstacle that even a superintelligent algorithm would find insuperable.

The problem lies in us humans. The oncologists of the future would be tempted by the same sense of familiarity that Kasparov experienced when watching AlphaZero. They will find flattering echoes of their own thinking in the algorithm's responses, even though it will have long since departed from any diagnostic principles that would still be intelligible to humans. The lingering human command of the subject would be an illusion.

This is a problem that arises when there are unbridgeable gaps between two subjects' cognitive powers. It is one thing for an academic philosopher to explain the essence of his job to his young children. They might have questions, but they would still have some semblance of understanding. It is quite another thing to perform the same task with the family dog. Like the children, she is a sentient being capable of processing information about her environment and pursuing her goals. But there is no explanation a human can offer her to make her understand what it means to be an academic philosopher.

We should expect Dr. Robot to run into the same problem. It may work out how to offer satisfying explanations to curious humans, but these will be circumscribed by the limits of our own cognitive powers. A dog might understand that its kibble comes from the human who pulls it miraculously out of the cupboard each morning. But the idea that the kibble must be purchased with money made by the human at his job as an academic philosopher would go well beyond her comprehension; and even if she could appreciate the contingencies you've identified, your explanation will still be meaningless, because it rests on unstated contextual knowledge that is simply unavailable to a dog.

At best, Dr. Robot might offer what *seems* like a very satisfying explanation. It might determine that the best way to fulfill its objective is to tell cancer patients something that isn't even technically true, but that has an intuitive quality of what comedian Stephen Colbert calls "truthiness." By offering us truthy explanations, Dr. Robot would be telling us what we need to hear without really conveying why it has made certain decisions.

What Would It Mean to Hand Off Control?

We are frequently told that we are going through a digital revolution. In that case, we should remember the eighteenth-century French royalist Jacques Mallet du Pan's famous observation that, "Like Saturn, the Revolution devours its children." The effects of digital computing—the transition from "atoms to bits"—have already permeated almost every corner of our economic, political, and social lives. The question now is whether we can avoid the fate of Georges Danton, who went from being the French Revolution's Minister of Justice in 1792 to becoming a victim of the guillotine in 1794.

As we've seen throughout the preceding chapters, we as a society are desperate to exert greater control over cancer, and we have alighted upon powerful machine learners to take us the rest of the way. But even if this hand-off were to go as planned, it doesn't follow that we would have gained more control; if anything, we may have let the disease slip from our grasp entirely.

This is not to conjure up dystopian scenarios like those envisioned by Swedish philosopher Nick Bostrom or the *Terminator* movie franchise, wherein an artificial superintelligence "becomes self-aware at 2:14 a.m. Eastern time, August 29th" and promptly turns humanity into batteries or game to be hunted. Rather, the loss of control could be much subtler. Indeed, if one looks around, one finds signs that it has already been happening.

The examples range from trivial cases of algorithms not doing what we want and expect of them ("why is the navigation app taking me *this* way?") to much more consequential events like democratic elections. In 2016, those in charge of the most powerful digital technologies of the day unwittingly played a significant role in the 2016 US election and the UK's Brexit referendum, and then were left aghast at the results.

In *Facebook: The Inside Story,* journalist Steven Levy recounts Facebook COO Sheryl Sandberg's experience on election night, 2016. She sent her daughter to bed "promising to wake her so she could witness history as the first woman president of the United States made her acceptance speech" (Levy 2020, p. 9). It wasn't to be, and in sharing this story later with Levy, Sandberg still gets "choked up" over Hillary Clinton's loss. One can imagine that somewhere deep in her subconscious, there is a gnawing awareness of the pivotal role her company played in helping the Trump campaign clinch victory by vanishingly small margins in a few swing states.

We know from leaked internal memos that Facebook's culpability in the 2016 election outcome has been the subject of intense debate within the company. One executive, Andrew "Boz" Bosworth, believes that Facebook was indeed "responsible for Donald Trump getting elected," (Bosworth 2020) because it gave the Trump campaign the tools it needed to reach its intended audience. By expressing shock and dismay at the effectiveness with which their platform has been used to advance ends antithetical to their own expectations, Facebook's leaders have essentially admitted to having a Frankenstein problem.

Of course, what Facebook executives are *really* thinking is impossible to say. As billionaires and multimillionaires, these individuals have little in common with most people on the planet, including most of their user base. How were they to know that introducing a largely unmoderated open platform in Myanmar would lead to an outpouring of hate speech, fake news, and genocide? In the colonial period, there were plenty of Europeans who sincerely wanted to understand the cultures and inner experiences of the societies they were subjugating. But, in the end, most settled for convenient fictions, such as British imperialists' claim that they were doing "primitive" societies a favor by "civilizing" them, even as they looted all of their wealth and natural resources.

Let's suppose that Sandberg is sincere. There can be little doubt that the author of *Lean In: Women, Work, and the Will to Lead* really wanted Clinton to win. If given a choice, she would most likely opt for a world in which her and Facebook's wealth was significantly diminished but Clinton had become president. Insofar as Facebook helped Trump get elected, it represents one of the many examples of humans losing control over the technology they have unleashed.

This loss of control is not the result of some hidden deviousness on the part of the algorithm or those who wrote it. It does not require us to cook up conspiracy theories, or to conjecture that the Facebook ecosystem is operating according to some secret stratagem beyond the "engagement"/profit-maximizing one that we already know about. Rather, the loss of control is a natural consequence of introducing machine learners into a highly complex, indeterminate system like a market economy—a place where "winners" are richly rewarded and "losers" are ruthlessly dispatched.

We can assume that Sandberg does not want to go down in history as one of the driving forces behind a company that derailed the election of America's first female president. And yet, as the powerful lead decision-maker behind Facebook's highly successful growth strategy, the 2016 election will feature prominently in her legacy. If history shows us anything, it is that even the most powerful people in the world are largely powerless in determining how they will be remembered.

Part of Facebook's problem is that its leaders have focused only on what can be reasonably predicted. Indeed, behavioral prediction *is* the business model. And in today's digital economy, where the network effect is king, it is reasonable to expect that the social-media platform that wins the race for scale will take home everything. It is this predictable outcome that Facebook has fixated on. Though its growth strategy has led it to become implicated in genocide in Myanmar, the Cambridge Analytica scandal in the United Kingdom, and many other controversies around the world, its market capitalization consistently climbed higher through all of these episodes. The near-term bet, at least, paid off.

Just as we should take Sandberg at her word, so, too, should we believe Facebook founder Mark Zuckerberg's original intention. "The thing that we are trying to do at Facebook," he explained early on in the company's history, "is just help people connect and communicate more efficiently" (Haupt 2021, p. 243). When he started out, he actually had no idea that this approach would generate the profits that it has. Who knew that there was so much money in encouraging strangers to "poke" each other online?

In recent years, Facebook has had to drop its original slogan—"move fast and break things"—and affirm repeatedly that it will "do better" when it comes to curbing hate speech and disinformation in an attempt to align its corporate brand with creating a better world (Haupt 2021). But even if these housekeeping efforts are sincerely pursued, we should expect them to fall short. There is simply too much money to be made from trafficking in sensationalism, and Facebook, after all, has shareholders to answer to.

But suppose that Facebook were to take seriously the adjusting of its algorithms to remove all fake news and hate speech from its platform. This would be very difficult and expensive, but Facebook is a well-capitalized company, and it already pays armies of content moderators in low-wage countries like the Philippines to ensure that pornographic images and terrorist beheadings are promptly deleted. There is presumably some sum of money that a $1.3 trillion company could commit to ensure that it is not providing a platform for content that is deleterious to democracy, let alone people's personal privacy or safety.

The obstacle to this seemingly obvious move becomes apparent when we consider the broader societal and economic environment in which Facebook operates. The decision to forgo these profits would be tantamount to inviting some less scrupulous competitor into the market. If Facebook were to adopt standards that diminish its ability to deliver "eyeballs" to advertisers, it would be shooting itself in the foot.

And so, for Facebook, the broader debate about whether it can and should try to exercise greater control over its platform is existential in nature. When Facebook rushes to acquire companies like Instagram, WhatsApp, and Oculus Rift—or when it simply copies other company's products—it is implicitly acknowledging that its time in the sun is limited. Politicians and the public can demand all they want of the company. But if Facebook's leaders have concluded that they can't both meet these demands fully and still survive, they will choose survival. They are not without agency, to be sure; but nor are they fully in control.

Out of Our Hands

If we hand off the Big Cs to the machines, one possibility, of course, is that we would lose oversight of problems against which we chart human progress. In this case, we might lead a cossetted existence free from the traumas of climate change, cancer, or coronaviruses.

But machine learners tend to expose the vagueness of the goals that we set for ourselves, and thus for them. We as humans think that we have a clear sense of what it might mean to inhabit a world without cancer. We have sought to bring about that world by declaring war on the disease and pursuing "the cure." Yet, as we saw in Chap. 3, there is considerable confusion in what it might mean to achieve these goals.

Moreover, even if we ourselves know what we mean when we speak of a cure for cancer at the species level—even if we accept that our thinking about the problem is metaphorical rather than logistical—it doesn't necessarily follow that we can convey our own understanding to a machine. Whereas our minds are comfortable with a degree of abstraction, an artificial intelligence would have to fill in gaps in reasoning that we take for granted. Billions of people have implicitly assented to Facebook's mission of making "the world more open and connected," only to find that they are surprised by many of the results.

Once we deploy machine learning to discern patterns in the vast trove of data on cancer, for example, we will have taken only the first step. What findings it will come up with, and what we will do with them, remains to be seen. Most likely, the process will expose elements of incoherence in our collective understanding of what it really means to create a world without cancer.

The problem, here, is not that a superintelligence would suddenly decide that the best way to fix the climate is to eradicate humans, thereby putting an end to anthropogenic carbon emissions. We are not alluding to the kind of existential risks popularized by Bostrom. Rather, the issue is that the Big Cs are complicated, not just scientifically but socially, politically, economically, and environmentally. Problems like the Big Cs lie at the nexus of many competing interests. There will always be biases embedded in the data, as is already the case for identifying potential skin cancers in populations with darker skin tones.

Past experience suggests that new technologies will be aimed first and foremost at the deepest markets. There is great interest in developing expensive new cancer drugs—like Avastin—to be sold to rich, well-insured people in the world's advanced economies. Meanwhile, there tends to be less attention on potential inexpensive interventions that could reduce the rate or mortality of cancer among poorer people. The divergence of interests between rich and poor is often concealed by the seemingly shared human commitment to curing cancer; but it would be laid bare when Dr. Robot takes over.

Likewise, once we renounce control over the Big Cs, gaps in our understanding of what would count as success against the disease will become more apparent. We have seen that cancer is an ineradicable feature of our essence as evolved beings that grow from a single cell, and that developing the disease becomes more likely as we age. To many people, merely keeping cancer at bay for long enough to reach a normal human lifespan would count as success. But there are also people with more exacting ambitions for radical life extension, and even if they are not in the majority, they will constantly be tugging on our collective expectations.

In addition to funding space exploration, Amazon founder Jeff Bezos is also investing in research to reverse the aging process. He knows that he can't take his billions with him when he goes, so why not commit some money to significantly delaying that fateful moment? Since we are imagining a future in which Dr. Robot becomes a reality, we also should allow for the possibility that anti-aging interventions will prove effective. As we've seen, Aubrey de Grey's vision of a millennial lifespan certainly belongs in the category of manifesto science, but it is not necessarily impossible.

In any case, significantly longer human lifespans would have direct implications for the War on Cancer. If you want to live to be 200, 500, or 1000 years old, you will need to demand much stronger therapies. Right now, being able to keep cancer at bay for 80 years could be interpreted as a final victory in the war. But in a future of longer lifespans, it will once again represent a failure. The war will have to be declared all over again, and past experience suggests that the treatments that follow will be ever-more expensive, available only to the few who can afford them.

How would a Dr. Robot sort through these value judgements? Would it be led by some quirk in the data or in its own programming to choose winners and losers? Over time, Facebook has increasingly drawn criticism for its role in haplessly elevating right-wing conspiracy theories like QAnon. What will be said about a medical algorithm that appears to have been empowered with life-and-death decision-making?

We face a choice quite unlike anything our species has considered before. One option is to try to expedite the development of machines to which we can hand off our biggest problems. Then, we could get on with leading the kind of cosseted lives Aldous Huxley lampooned in his 1932 novel *Brave New World*.

Alternatively, we could choose to keep distinctive human problems in human hands (Agar 2019). That might lead to more casualties along the way, but it also would retain our ability to question and challenge societal priorities. When we hand things off to algorithms whose reasoning we cannot understand, we will also be surrendering a degree of ethical agency. What happens when Dr. Robot tells us that the poor are indeed at a higher risk of cancer than the rich, but cannot tell us how it knows this? What if its recommended treatment is to reduce all forms of socioeconomic inequality? Will we then deploy a separate AI to tackle that problem (or ask AlphaInfinity to take care of both)?

Susan Sontag challenged the metaphor of the War on Cancer on the grounds that it merely recapitulated the moral imperfections of the culture. She hoped for a future in which cancer would be a simple scientific problem. The good thing about metaphors is that they are creations of human minds. We can challenge a dominant metaphor in ways that we would not be able to challenge the machine, once we have relinquished our tenuous grip on the emperor of all maladies.

References

Adams, Henry. 1996. *The Education of Henry Adams*. New York: The Modern Library.
Agar, Nicholas. 2019. *How to be Human in the Digital Economy*. MIT Press.
Bosworth, Andrew. 2020. *Facebook post*. 7 January. https://www.facebook.com/boz/posts/10111288357877121. Accessed 17 Mar 2020.
Brynjolfsson, Erik, and Andrew McAfee. 2017. *Machine, Platform, Crowd: Harnessing Our Digital Future*. W.W. Norton.
Burns, D.M. 2000. Cigarette smoking among the elderly: Disease consequences and the benefits of cessation. *American Journal of Health Promotion* 14 (6): 357–361.
Contagion. 1997. *Directed by S. Soderbergh [Film]*. Warner Bros. Pictures.

Crispo, A., P. Brennan, K.H. Jöckel, A. Schaffrath-Rosario, H.E. Wichmann, F. Nyberg, L. Simonato, F. Merletti, F. Forastiere, P. Boffetta, and S. Darby. 2004. The cumulative risk of lung cancer among current, ex-and never-smokers in European men. *British Journal of Cancer* 91 (7): 1280–1286.

De Waal, Alex. 2021. *New Pandemics, Old Politics*. Polity.

Domingos, Pedro. 2015. *The Master Algorithm: How the Quest for the Ultimate Learning Machine Will Remake Our World*. Allen Lane.

Dougherty, Edward R. 2016. *The Evolution of Scientific Knowledge: From Certainty to Uncertainty*. SPIE Press.

Editorial. 2021. Artificial Intelligence in Structural Biology Is Here to Stay. *Nature*, 595, 625–626, 27 July. doi:https://doi.org/10.1038/d41586-021-02037-0.

Haseltine, William A. 2021. Can We End the Pandemic? *Project Syndicate*. 4 June.

Haupt, Joachim. 2021. Facebook futures: Mark Zuckerberg's discursive construction of a better world. *New Media & Society* 23 (2): 237–257.

Knight, Will. 2020. Defeated Chess Champ Garry Kasparov Has Made Peace With AI. *Wired*. 21 February.

Levy, Steven. 2020. *Facebook: The Inside Story*. Blue Rider Press.

Mossman, Kenneth L. 2014. *The Complexity Paradox: The More Answers We find, the More Questions We Have*. Oxford University Press.

Russell, Stuart. 2019. *Human Compatible: Artificial Intelligence and the Problem of Control*. Viking.

Silver, David, Thomas Hubert, Julian Schrittwieser, Ioannis Antonoglou, Matthew Lai, Arthur Guez, Marc Lanctot, Laurent Sifre, Dharshan Kumaran, Thore Graepel, Timothy Lillicrap, Karen Simonyan, and Demis Hassabis. 2018. Mastering Chess and Shogi by Self-Play with a General Reinforcement Learning Algorithm. *Science* 362 (6419): 1140–1144. https://doi.org/10.1126/science.aar6404.

Solon, Olivia. 2017. Sorry, Y'All—Humanity's Nearing an Upgrade to Irrelevance. *Wired*. 21 February.

Strogatz, Steven. 2018. One Giant Step for a Chess-Playing Machine. *The New York Times.*. 26 December.

The Economist. 2013. Difference Engine: Crash Program? *The Economist.* 26 August.

Turing, Alan. 1950. Computing Machinery and Intelligence. *Mind* LIX (236) October: 433–460. https://doi.org/10.1093/mind/LIX.236.433.

Chapter 8
Waiting for the Techno Rapture

Contents

Overly lofty ambitions built on an extrapolation from present knowledge are an abiding feature of modernity. In the mid-nineteenth century, the consensus among radical reformers around the world was that technologies born of the Industrial Revolution would allow for the creation of an entirely new society. These aspirations were embodied in The Crystal Palace, a magnificent cast-iron and glass structure that had been unveiled at the Great Exhibition in London in 1851. Capturing the mood of the era, the Russian intellectual Nikolai Chernyshevsky saw that architectural feat as merely the beginning.

In his 1863 prison novel, *What Is to Be Done?*, Chernyshevsky envisions a future in which "life is spacious and abundant," allowing for everyone to live "just as he pleases." The deserts of the Middle East will have been converted into fertile farmland, recreating the biblical "land of milk and honey." People will have figured out how to transport "clay on their mighty machines to bind the sand," and to follow up these efforts with irrigation canals." The key to this success will have been no great secret. "People merely became more intelligent and began to turn to their own advantage the tremendous means and resources that once had been wasted or used counterproductively." Prefiguring today's techno-optimists, Chernyshevsky tells us that, "Once they began to understand, it was not hard for them to make progress" (Chernyshevsky 1863, p. 374).

Though it was hardly a great artistic achievement, *What Is to Be Done?* became one of the most important political artifacts of the modern era. It, "far more than

© The Author(s), under exclusive license to Springer Nature Switzerland AG 2024
N. Agar et al., *How to Think about Progress*, Library of Ethics and Applied
Philosophy 42, https://doi.org/10.1007/978-3-031-68938-3_8

Marx's *Capital*," the literary scholar Joseph Frank contends, "supplied the emotional dynamic that eventually went to make the Russian Revolution" (Chernyshevsky 1863, p. 1). Indeed, Vladimir Lenin is said to have read the book five times before adorning one of his own famous pamphlets with the same title.

Of course, today's evangelists of technological progress are not reprising the Bolshevik project. Yet they do tend to exhibit a similar form of secular faith. Their own manifestos lead them to believe that the conceivable is therefore imminent. But, of course, timing is everything. Launched in 2009, the Sahara Forest Project, in partnership with major fertilizer companies, promises to "green" parts of the desert in Qatar through the use of already available "environmental technologies" (Herben 2014). It would seem that Chernyshevsky wasn't wrong about the material feasibility of his vision. He was just far too early.

In thinking about today's biggest challenges, we would do well to remember the Copernican principle, which holds that we are far more likely to occupy a non-special place in time (or in the universe) than a special one. Like Nixon and all those since who have foreseen an imminent cure for cancer, we may *feel* that we are 90% of the way there. But it is just as likely that we are actually only 10% of the way—or 30% or 60%. As the astrophysicist J. Richard Gott III pointed out in a famous 1993 paper, once we accept that our own position in time is random, we also must accept that it probably isn't special (Gott 1993). Why should we of all generations be the ones to triumph over disease or conquer new planets?

In the 1820s, Johann Wolfgang von Goethe, nearing the end of his life, weighed in on a debate about human progress that has been evolving at least since the time of Xenophanes in the late sixth century BCE. Humanity may develop for millions of years, Goethe told the German poet Johann Peter Eckermann, "But let humanity last as long as it will, there will always be hindrances in its way, and all kinds of distress, to make it develop its powers. Men will become more clever and discerning, but not better nor happier nor more energetic, at least except for limited periods." Goethe also warned of what we have called the horizon bias. "The world will not reach its goal so quickly as we think and wish," he observed. "The retarding demons are always there, intervening and resisting at every point, so that, though there is an advance on the whole, it is very slow. Live longer and you will find that I am right" (Eckermann 1930, p. 275).

Many other voices of the time were not nearly as nuanced in their views. The first half of the nineteenth century was when grand theorizing about "Progress" reached a fever pitch, becoming the defining intellectual force behind modern civilization itself. Building on the *Encyclopédists'* sketches of universal history in the previous century, radical thinkers like the positivist Auguste Comte described a "law of human progress" that can be used not only to explain all of history up to the present, but also to predict the general course of human development in the future.

Comte, who believed that human societies ultimately follow the same kinds of deterministic laws as the physical world, was simply following to its logical conclusion a certain strain of Enlightenment thought about progress. In the seventeenth century, René Descartes and Francis Bacon had both broken forcefully with the ancient intellectual tradition, arguing that nothing can be taken for granted. Rather than reflexively honoring the Greeks, the Romans, or the Church, each man believed that much of the knowledge that had been handed down through the generations was a mix of unfounded metaphysics and obscurantist cant.

For his part, Bacon described the great philosophers of the past, from Aristotle to St. Augustine, as spiders, spinning webs from their own minds with no reference to reality; only by combining reason with experimentation—like honeybees—could genuine knowledge be acquired. Likewise, Descartes showed how the proper application of human reason could—indeed, must—cast doubt on all received wisdom. In devising a new methodology of thought, he assumed that the natural phenomena available to our sense perceptions were beholden to constant laws.

It was on this intellectual bedrock that the scientific and industrial revolutions would emerge. Gradually, people started to recognize that insofar as anything can truly be known about the world, such knowledge can be acquired through experimental science and human reason. And once gained, it can be put to practical use.

The Modern Creed

Yet far from representing a sharp break from the superstitious dogma of the past, the Cartesian and Baconian streams combined with another expanding intellectual tributary: Puritanism. The Puritans of the seventeenth century saw the new science not as a threat to their beliefs but as a gift from Heaven. It was through science and technology that God's kingdom on Earth would be achieved. In the millenarian worldview of the time, Providence and Progress were more or less synonymous.

During the eighteenth and nineteenth centuries, the role of a divine interlocutor was increasingly made redundant by substitutes such as the "invisible hand," the "cunning of reason," dialectical materialism, and other secularized prime movers. Human progress could be explained without recourse to superstition, but the basic idea still required a central mechanism. As the eighteenth-century *Encyclopédist* Anne Robert Jacques Turgot averred, "progress leads to further progress" as a matter of course (Turgot and Gordon 1750, p. 323).

An early technological determinist, Turgot believed that as long as humanity's store of shared knowledge could be preserved through writing and printing, it could only ever increase and become more useful. Like many other theorists of progress since St. Augustine, he analogized the "human race" to a single individual who has for centuries been gradually advancing since "its infancy," albeit with no possibility for old age in its future. As we saw in Chapter Six, this collective-mind metaphor has begun to run into limits such as the "burden of knowledge" problem and the drive toward ever deeper, more fragmented specialization in academia.

In any case, even as the modern doctrine of progress became increasingly secular, it remained faith-based to the core. "Fundamental to the idea," the sociologist Robert Nisbet explains in his *History of the Idea of Progress*, "is faith in the value of human knowledge, the kind of knowledge that is contained in the sciences and the practical arts, and faith also in the capacity of such knowledge to lift humanity to ever-higher levels of human life" (Nisbet 1980, p. 128).

It is this particular faith to which Goethe was referring 200 years ago, and that still underpins so much manifesto science today. Where once the doctrine of progress inspired socialist and Marxist social reform projects, it has now been taken up by libertarians and other techno-optimists who believe that if people are free to do as they please, their organic acts of creation will automatically "make the world a better place" (CATO 2014).

Like the Puritans, we today are not just hopeful but impatient for timely technological solutions to all of our most vexing problems. In America, despite certain subcultures of anti-science crankery, three-quarters of Pew survey respondents believe that science has a "mostly positive impact on society," and 82% are confident that "scientific developments will continue to improve lives" (Thigpen and Funk 2019).

Consider climate change and related issues of environmental degradation. Few other problems have invited as much technological-deterministic thinking. In his 2018 paean to progress, *Enlightenment Now*, Steven Pinker assures us that "environmental problems, like other problems, are solvable, given the right knowledge." The data show, he insists, that many are already being solved as a byproduct of technological progress. "Something in the nature of technology, particularly information technology," Pinker claims, "works to decouple human flourishing from the exploitation of physical stuff." Thanks to all of the advances of the past few decades, humanity has become "more resilient to natural and human-made threats: disease outbreaks don't become pandemics … populations are better protected against storms, floods, and droughts" (Pinker 2018, p. 136, 142).

The arrival of the COVID-19 pandemic just 2 years after publication of Pinker's book underscores the tenuousness of such claims, as do projections of a massive increase in climate refugees in the coming years (if they are anywhere near accurate) (Rigaud et al. 2018, p. 2). Indeed, any such claim about the inherent "nature of technology" ultimately falls into the realm of metaphysics. While it is certainly true that efficiency-boosting technology can shrink humanity's environmental footprint, it also can have the exact opposite effect, by enabling people to consume even more of something than they previously would have (an economic phenomenon known as Jevon's Paradox).

More to the point, the repeated disappointments in the War on Cancer should lead us to question our unconditional faith in the capacity of technology to deliver "just-in-time" solutions for problems as massive and consequential as climate change. Once we realize that the prevailing narrative of progress rests ultimately on faith-based claims about the trajectory of our own civilization and inventive capacities, we ought to be more prudent in betting on our prospects for future prosperity and survival.

Imperfect But Necessary Metaphors

When the idea of progress becomes a grand historical theory or an object of meta-physical speculation, it functions as a reductionist metaphor. This is part of a larger pattern. When we are not handing off truly hard problems to AI, we tend to tackle them with a much older tool: metaphoric thinking.

Howard Moss, the long-time poetry editor of the *New Yorker*, once wrote that, "The power of metaphor is not merely descriptive but psychological; the link between two things we were not aware of is revealed to us. Far-fetched it may be, even bizarre; we know instantly, though, whether it rings true" (Moss 1962, p. 38). But ringing true is merely a necessary condition; it is not sufficient for making a metaphor useful.

Some metaphors can do more harm than good by simplifying complex phenomena too much. Comparing the human body to a piece of mechanical hardware is a common example. A biological phenomenon like the immune system is orders of magnitude more complex than an automobile; nonetheless, some physicians have found this particular analogy helpful when speaking to patients or the public (Pardoll 2016).

Equally, our metaphorical wars have drawn many critics. We are constantly declaring war on problems like disease, climate change, poverty, inequality, and terrorism without first defining the terms of victory. As Alex de Waal warns, a "'war on disease' is not a harmless metaphor," because it "suppresses critical thinking" and suggests a false "comforting ending" (De Waal 2021, p. 2).

Similarly, in the context of cancer, British oncologist Paul Workman discourages the use of "aggressive military language" (Workman 2019). Many of his patients don't want to think of themselves as soldiers in a war. Likewise, Kate Pickert writes, "Despite the familiar archetype of a breast cancer patient who battles her disease, I never felt like I was at war" (Pickert 2019, p. 4). These objections must be respected. No one can or should dare to tell a cancer patient how she should think about her own condition.

Moreover, this flaw is not limited to the War on Cancer. It also has a glaring presence in yet another ongoing metaphorical war: the War on Drugs. In the 2000 film *Traffic*, the character Robert Wakefield, a hardline judge played by Michael Douglas, discovers that his daughter is an addict who has been sucked into prostitution. Long an ardent supporter of the War on Drugs, he eventually comes to see the entire project as misguided. "If there is a war on drugs, then many of our family members are the enemy," he admits. "And I don't know how you wage war on your own family."

On the other hand, there are others for whom the war metaphor offers inspiration. Some patients and their families are motivated by the idea that they are on the front lines of a war. Some will respond to the type of rhetoric that we use to inspire our young people to fight for their countries—to find meaning in something larger than themselves. Nothing could be more personal than one's own experience with cancer; but in the War on Cancer, one is guaranteed to have company. In fact, there is even an app for it. The "storytelling platform" War On Cancer seeks to "radically improve the mental health of everyone affected by cancer," by reminding them that they are "never alone" (WarOnCancer 2020).

Nonetheless, even if it turned out that the war metaphor enjoyed majority support, it still could have knock-on effects that should be acknowledged and accounted for. The psychologists David J. Hauser and Norbert Schwarz have studied the issue and concluded that "battle metaphors increase fatalistic beliefs about cancer prevention," and thus "can influence the health beliefs of nonpatients in ways that may make them less willing to enact healthy behaviors" (Hauser and Schwarz 2019). In other words, far from offering motivation, the War on Cancer could simply deepen one's despair or resignation. While the soldiers in a war might feel a sense of agency, the civilians—the collateral damage—tend to have the opposite experience. And if cancer is such an insurmountable and inevitable affliction, why not have that cigarette?

Even so, it is easy to see why people cling to the war metaphor. It will not resonate with all doctors, researchers, or policymakers, and it should certainly be used with care at the bedside; but it still may be the best conceptual framework we have for setting public policy priorities.

The fact that the war metaphor has been overused in our responses to terrorism and drugs is a testament to its power. Like many overused clichés, war has the advantage of universal clarity. We all know intuitively what it signifies, and what it signals for the future. As soon as the realities of the COVID-19 pandemic had dawned on the Western public, calls for a "wartime mobilization" became commonplace.

A declaration of war establishes a collective cause, taps into an underlying sense of duty, and commits policymakers to a certain path. It's unfortunate that humans are so easily stirred to collective action by talk of war, but the fact is that we are. If we really could re-engineer our fundamental beings, we might consider changing that feature of human nature. For now, though, we should work with the motivational tools that are available to us.

If nothing else, the War on Cancer has ensured that cancer is never one of those neglected diseases that one hears about only when it suddenly afflicts some celebrity or public figure. In the US alone, the war has mobilized at least $500 billion in research funding, allowing us to be confident that the future will at least bring more effective treatments. If Biden and Trump had limited themselves to promising that better therapies would become available during their time in office, they would have spoken truthfully. It would be much better to receive a cancer diagnosis at the end of those 4 years than at the beginning.

The War on Cancer also serves as a clear statement of priority. There are countless needs that modern societies could spend money on. We could fund artists to beautify our cities, launch programs to improve citizens' physical fitness, or dump billions of dollars into recruiting a soccer team capable of winning the World Cup. The fact that we have committed to a war on cancer means that developing improved cancer therapies takes priority over many of these other potential collective goods.

Moreover, what we collectively prioritize is limited not only by resources, but also by attention. In reference to collective efforts to address climate change, the psychologist Elke U. Weber speaks of a "finite pool of worry" (Weber 2006, p. 114). We have only so much time and mental capacity with which to focus on issues that might warrant our attention. An entire society's pool of worry will be much larger than what we can imagine as individuals; but it is still finite. And "unlike money or other material resources," Weber points out, attention can be neither saved nor borrowed (Weber 2010, p, 334).

Attention is, however, fungible. Worries about climate change could quickly transform into worries about an apocalyptic asteroid strike, were one suddenly to show up on the radar. But attention's fungibility also creates challenges. For those who understand climate change to be a serious problem that warrants immediate attention, the task is not just to get people to acknowledge the threat. The public also needs to be persuaded to allocate sufficient quantities of worry from the collective's finite pool, and it needs to do so in a sustained fashion. Not surprisingly, proponents of a Green New Deal have tapped into the power of the war metaphor: though we've yet to witness a formal declaration of war, climate change is now routinely said to demand a World War II-scale mobilization (Romm 2018).

Sometimes, "economies of scope" are available when allocating finite attention and resources. The Green New Deal has been billed not just as a climate policy, but also as a means of reducing inequality and serving social-justice objectives. When limiting global warming, we also may be engaging in pandemic mitigation, by keeping pathogens currently locked under permafrost where they are (Fox-Skelly 2017). In these cases, our attention "budgets" can be shared.

But double-dipping won't always be possible, which is why we need clear priorities. And an agenda item far enough down the list of priorities may as well not be on the list at all. Politicians who agree in principle on the need to reduce carbon emissions may end up not doing much about it, because they may be convinced that facilitating (non-green) economic growth is more important. Indeed, for some politicians and commentators, this argument has become the new climate denialism—discounting not the science but the moral urgency of climate change.

The War on Cancer has undeniably pushed procurement of better cancer therapies up the list of good things that we can collectively achieve. It has put cancer treatment ahead of a host of other meritorious ends. The climate crisis may well belong higher on that list than the War on Cancer. It may even be an exaggeration to say that there is a "cancer crisis" in the same way that there is a climate crisis. (It is not as though cancer diagnoses and deaths are skyrocketing.) But we are where we are. The War on Cancer may not have gone well over these past 50 years, but that doesn't mean we should have directed our attention elsewhere.

Besides, there is that other finite resource to consider: money. As the historian Walter Scheidel shows in his wide-ranging history of inequality, *The Great Leveler*, wars—especially modern mass-mobilization wars—are one of the few occasions upon which the haves will commit wealth and resources to a collective cause (Scheidel 2018). In times of peace, equality tends to remain merely an ideal, and the wallets of the rich remain closed. But Scheidel finds that a broad-based sense of duty comes into play during wars and other societal crises that make everyone feel as though they are in it together.

During World War II, for example, British elites proved more willing than in the past to sacrifice some of their own wealth for the cause of victory. Certainly, self-interest was involved in their change of attitude. But they also knew that young, working-class people were laying down their lives to defeat fascism, and that waging the good fight would require weapons and other supplies. Similarly, after the terrorist attacks of September 11, 2001, American taxpayers had few qualms with spending hundreds of billions of dollars in the War on Terror, whereas if that money had been earmarked for poverty relief, it would have invited howls of protest from some significant share of the electorate.

In any case, the United States can't launch a War on Poverty now because it already did so more than 50 years ago. In 1964, President Lyndon B. Johnson declared "unconditional war on poverty in America." Unlike Nixon, Johnson acknowledged from the start that, "It will not be a short or easy struggle, no single weapon or strategy will suffice, but we shall not rest until that war is won."

Was the War on Poverty won? Like most policy questions in America, the answer depends on which party you ask. Marking the 50th anniversary of Johnson's declaration in 2014, those on the right described the war as a failure. After "U.S. taxpayers have spent over $22 trillion on anti-poverty programs," noted Rachel Sheffield and Robert Rector of the conservative Heritage Foundation in 2014, "progress against poverty, as measured by the U.S. Census Bureau, has been minimal, and in terms of President Johnson's main goal of reducing the 'causes' rather than the mere 'consequences' of poverty, the War on Poverty has failed completely" (Sheffield and Rector 2014, p. 1).

By contrast, scholars at the left-leaning Center for American Progress argued that, "While many of the programs that emerged from this national commitment are now taken for granted, the nation would be unrecognizable to most Americans if they had never been enacted." Thus, to label the effort a failure, "is to say that the creation of Medicare and Head Start, enactment of civil rights legislation, and investments in education that have enabled millions of students to go to college are a failure" (Boteach et al. 2014).

Such differences of opinion are perhaps inevitable when one goes to metaphorical war, especially when the very definition of victory is contested. Like the War on Cancer, the War on Poverty allows for many individual "moonshots"—concrete social programs, new cultural norms—that may or may not change one's basic calculus about the broader effort. And like cancer, poverty is the result of highly complex, dynamic processes. Saving one community—or even one generation—from poverty is not the same thing as *ending* poverty. Around the world, tens of millions of people were lifted out of poverty during the first two decades of the twenty-first century. But following the global recession brought on by the COVID-19 pandemic, the World Bank warned that a significant share of these gains were thrown into reverse (Abi-Habib 2020).

Likewise, in the US in recent years, economists have increasingly warned that social mobility is falling: that fewer and fewer younger Americans can expect to earn as much as their parents did (Cook 2019). Such historic reversals point to the problem of assessing progress in any broad social effort. In the decades following

World War II, American middle-class incomes grew steadily. A growing share of the population went on to earn a college degree, to own a home, and to check off other boxes of the "American Dream." Though these gains did not accrue in equal proportion to African-Americans and other disadvantaged cohorts, they nonetheless constituted measurable progress. The post-war increase in American standards of living was real. Millions of Americans experienced it firsthand, and it is now a matter of historical fact.

The post-war Golden Age will forever remain a fixture in US history books. The question, though, is whether that permanent record qualifies as a permanent achievement. On the one hand, the post-war period furnished us with untold, universally affordable modern wonders that we now take for granted (refrigerators, televisions, washing machines, and the like). It also arguably created the conditions for the social policies of the 1960s, including not just the War on Poverty but also the Civil Rights Act.

On the other hand, despite all of the gains achieved during the post-war period, American life expectancy has plateaued or even fallen for some population cohorts. Median household incomes have stagnated (Case and Deaton 2020). Inequality, by many measures, has steadily increased (Desilver 2018). And widespread social and economic disparities remain an endemic problem (Elias and Paradies 2021). Clearly, "progress" is not automatically bequeathed to the next generation. Each generation must fight the battle anew, and oftentimes that task will involve correcting past errors, or even abandoning previous definitions of progress. There will always be new challenges to equality, justice, and material abundance, just as there will always be new potential carcinogens born of chemical synthesis and the other engines of material growth and development.

Such is the nature of what the sociologist Zygmunt Bauman calls "liquid modernity." The common denominator in all forms of modern life, he argues, is a deep-seated "fragility, temporariness, vulnerability and inclination to constant change." Whereas those in the early twentieth century thought that "'to be modern' meant to chase 'the final state of perfection,'" he writes, "now it means an infinity of improvement, with no 'final state' in sight and none desired" (Bauman 2012, p. 82). In the wars on both cancer and poverty, it is not just the enemy that is fluid and ever-changing. So, too, are we.

Indeed, with respect to cancer, the idea of "the cure" is itself a metaphor. Though many fortunate individuals are regularly cured of their particular cancer, we as a species are not. The individual's experience cannot be what we mean when we speak of finding *the* cure. Eradication or a truly "universal cure" would fit the bill; but, as we've seen, neither is really an option. As a metaphor, though, the idea of the cure offers hope, if not a guarantee, for clarity—for something that rings true. What is really meant is some future scenario where cancer will no longer weigh more heavily on our minds—or health-care budgets—than other common problems.

As Susan Sontag pointed out, metaphors can serve as a kind of acid test for progress. Noting that the word cancer itself is frequently used as a metaphor (as in, "so-and-so is a cancer on society"), she suggested that we will know that we have overcome the disease when it has lost its metaphorical power. "It may then be possible to compare something to a cancer without implying either a fatalistic diagnosis or a rousing call to fight by any means whatever a lethal, insidious enemy. Then perhaps it will be morally permissible, as it is not now, to use cancer as a metaphor" (Sontag 1978, p. 728).

By the same token, we may know that we have overcome problems like COVID-19 or climate change when we no longer feel the need to wage "wars" on them. Insofar as these metaphors "ring true," they will have their uses.

Perspectives on Progress

By contrast, metaphors of progress that treat humanity as a single unit moving toward or from some point in time are problematic. These tend to contribute more confusion than clarity. When one hears such claims, one immediately must ask what progress even is, and by what criteria it should be assessed.

To be sure, many definitions have been put forward over time. In Nisbet's overview, progress is simply "the idea that civilization has advanced in the past, is now advancing, and in all likelihood will continue to advance in the foreseeable future" (Nisbet 1980, p. 4–5). To Pinker, it is "the idea that the world is better than it was and can get better still" (Pinker 2018, p. 39). Both definitions come close to the one offered by the early twentieth-century historian J.B. Bury: "The idea means that civilization has moved, is moving, and will move in a desirable direction" (Bury 1920, p. 1).

But each of these definitions merely begs the question. If one were to ask an Islamic State leader in 2013, following the capture of Raqqa, whether his civilization was advancing in a desirable direction toward a better world, he would have answered in the affirmative. This is an extreme example, to be sure. We raise it not to suggest that everything is relative, but rather to show that progress is not nearly as rigorous an idea as its common usage implies. More often than not, it comprises only means (average lifespans, incomes, technological conveniences) but not ends (individual or collective meaning, the good life). Until such terms are clarified, we cannot possibly hope to manage our collective expectations about the future.

Moreover, even if everyone were to agree about what constitutes progress, no one can actually know what actions and innovations will prove to be progressive, as opposed to inconsequential or even regressive. Immediately following Trump's election in November 2016, Mark Zuckerberg wrote on Facebook, "We are all blessed to have the ability to make the world better, and we have the responsibility to do it" (Acton 2016). Almost 2 years later to the date, Facebook admitted that it had been used to "foment division and incite offline violence" in Myanmar—violence that the United Nations High Commissioner for Human Rights has described as genocide (Warofka 2018; UNHRC 2018). Or, to use an example from the field of cancer medicine, there is strong evidence to suggest that the overuse of mammograms has merely increased health-care costs without providing proportionate medical benefits (Miller et al. 2014).

Whether these episodes can be slotted into a grander narrative about progress is a matter of interpretation. In 1945, the United States demonstrated the destructive power of the atom bomb over Hiroshima and Nagasaki, inaugurating a new era of nuclear deterrence. Did that actually limit the possibility for wars between countries and thereby yield progress? Such questions can be debated endlessly, but much will simply come down to perspective. Those with an optimistic outlook are like the *Encyclopédists*, celebrating humanity's steady climb upward from its primitive infancy. Some, like Turgot, even see humanity's errors as being progressive, trusting that we will always learn from our mistakes.

But if one insists on imputing a moral arc to history, it is all too easy to invert the picture, as did Walter Benjamin on the eve of World War II:

> A [Paul] Klee painting named 'Angelus Novus' shows an angel looking as though he is about to move away from something he is fixedly contemplating. His eyes are staring, his mouth is open, his wings are spread. This is how one pictures the angel of history. His face is turned toward the past. Where we perceive a chain of events, he sees one single catastrophe which keeps piling wreckage upon wreckage and hurls it in front of his feet. The angel would like to stay, awaken the dead, and make whole what has been smashed. But a storm is blowing from Paradise; it has got caught in his wings with such violence that the angel can no longer close them. This storm irresistibly propels him into the future to which his back is turned, while the pile of debris before him grows skyward. This storm is what we call progress. (Benjamin 1955, p. 201)

Rather than progress, the best that can be offered as a law of human development is the basic observation that humankind *accumulates*. We tend to amass more information, more material stuff, and more social institutions with the passage of time. While we chalk up ever more achievements, cures, marvels, and historical feats, so too do we suffer setbacks, tragedies, and episodes of sharp retrogression.

Even when we can be confident that we will know how to solve a particular problem, we cannot know when the needed breakthroughs will occur. Chernyshevky wasn't necessarily wrong in the nineteenth century when he anticipated that deserts would be made into verdant, cultivated land; but he was certainly premature. Like us today when it comes to cancer or climate change, he had an abiding faith in "just-in-time" progress.

Ultimately, though, claims that treat progress as a metaphysical feature of history or an automatic byproduct of human ingenuity will run into contradictions. Whether we as a species are "better off" than our ancient ancestors, whether we are marching toward a still-better future—these are not actionable questions. If one could travel back in time to the shit-filled streets of Pompeii, one would doubtless find the amenities to be sub-standard. But that would merely reflect one's own twenty-first-century acclimation. It would say nothing about the average Pompeii native's sense of the present or aspirations for the future.

Similarly, when Pinker, touting the wonders of modern entertainment offerings, writes that, "It's hard for us to reconstruct the gnawing boredom of the isolated rural households of yesteryear," he is absolutely correct, albeit not for the reasons he suggests. The scenario *is* hard—indeed, impossible—to reconstruct, because we are being asked to believe, preposterously, that the vast majority of human beings who have ever lived suffered from what we in our uniquely internet-addled condition would call "gnawing boredom" (Pinker 2018, p. 260).

For all that we know, some people of yesteryear would look on us with the same degree of condescending pity with which evangelists of progress look upon past peoples. After all, ours is the generation suffering from crises of "post-truth" politics, chronic loneliness, and "deaths of despair" (Hertz 2021; Case and Deaton 2020).

Progressing While Cancering

The doctrine of progress that guides our sense of civilizational purpose can be distinguished from narrower, more colloquial forms of advancement, such as improved treatments for cancer, better space-faring gear, or reductions in greenhouse-gas emissions. This book has offered reasons for why one should be skeptical of promises for "just-in-time" progress against problems like cancer and climate change.

As Goethe saw long ago, there will always be unexpected hurdles standing in the way of even the most informed among us; and hedonic normalization will limit future generations' ability to feel substantially "better" than we do about our prospects as biological "cancering" beings.

We can predict that there will be a persistent reluctance to confer the honorific "cure" label on any advance against cancer, no matter how impressive it seems at first. We should not expect a perfect or "universal cure" that immediately restores every patient to a state indistinguishable from that of one who has never had cancer. Rather, we should assume that there will continue to be differences between the experiences of cancer patients and everyone else, and that these differences will continue to matter to each given generation in its own time. A time when we can declare victory in the War on Cancer without inviting the resentment or mockery of our descendants may never come.

Or, perhaps changes in our circumstances will lead us to worry less about cancer, because there's simply so much more to worry about. A gradual shift in our allocation of "finite worry" may lead us to pronounce victory in retrospect. As rising seas flood our coastal cities, people may stop thinking as much about cancer, even though it will persist. A cancer diagnosis of course would still be bad news for those who get it; but it might claim a smaller share of humanity's collective attention.

At that point, our descendants might busy themselves waging metaphorical wars against desertification and city-wrecking storms, wondering why their forebears had chosen to expend so much time, energy, and resources eking out diminishing returns in a war against an ineliminable part of the human condition.

A constant challenge will be to reconcile the attitudes associated with progress *toward* and progress *from* various benchmarks in humanity's struggle against cancer. In between, we can try to focus on a kind of non-prepositional progress that still captures our collective sense of achievement following objective technological advances against the disease. While ideas about progress *since* the time of Imhotep and progress *toward* some aspirational horizon are both valid perspectives for assessing the current state of play, the former tends to be more upbeat than the latter. We know that we've come a long way, but we cannot know what we have to look forward to, or how much farther we have to go.

Still, we can reconcile these perspectives by converging on an emotionally satisfying message of *simple progress* based on existing technologies, rather than yet-to-arrive breakthroughs. Though we should not expect sudden, sweeping solutions to major challenges, that doesn't mean we need to abide a sense of collective despair.

We have ample inductive evidence to believe that there will continue to be gradual improvements in the technologies for fighting climate change or treating cancer and other diseases. It is safe to say that even if we cannot win the war anytime soon, nor can we lose it (collectively, at least). Though we may choose to hand things over to the machine, there is no conceivable future in which we will have surrendered to the emperor of all maladies.

One implication of this perspective is that we will continue to regard our metaphorical "wars" as spending priorities. True, we could just end the effort against cancer and pronounce ourselves content with existing therapies and the notable progress that has already been made through prevention (tobacco cessation, the human papilloma virus vaccine, sunscreen, etc.). After all, if we're doomed to experience ongoing perpetual disappointment, what's the point of expending so much time, resources, and energy on this particular challenge? One byproduct of societal aging in the advanced economies that are the most focused on cancer is a quietly emerging dementia epidemic. Why not shift some of the focus to this other category of disease for which there is no known "cure"?

We will indeed have to address these other diseases of aging, lest progress against one be offset by the increasing prevalence of the other. In each case, we will need to stop longing for breakthroughs and instead settle for simple progress, which would include any advance that leads to therapies that are demonstrably superior to existing treatments.

Simple progress is sufficient to ensure that tomorrow's therapies are better than today's, but it does not imply a future in which all informed witnesses would agree that the War on Cancer and all other diseases had been won. It does not hold out hope that someone somewhere will deliver a fundamental breakthrough, further confirming humankind's primacy over nature.

Simple progress also should suffice to earn the gratitude of many people who are diagnosed with cancer. These people already make regular appearances in contemporary accounts of progress against the disease. Charles Graber's 2018 book, *The Breakthrough: Immunotherapy and the Race to Cure Cancer*, for example, offers many riveting anecdotes about individuals with terminal diagnoses who have been saved by bold new experimental therapies (Graeber 2018). Owing to these forms of coverage in the popular press, many people without cancer have come to see immunotherapy as the weapon that will deliver victory in the War on Cancer.

But these stories rely on a familiar survival-against-the-odds narrative, like the one about the child who falls through a frozen pond and survives; or the one with the pilot who makes a successful emergency landing in the Hudson River. We love these stories for their feats of human resilience and courage, not because they augur significant advances in the treatment of hypothermia or in aircraft safety.

Progress Is as Progress Does

Still, hopeful stories are not useless by any means. It is important for the "war effort" that we retain a collective sense of progress, lest public funding and support diminish. The public will need to understand that progress will almost certainly become even more difficult and require even more funding with the passage of time. It seems to cost more to pursue new breakthroughs now than it did in the 1950s, because we have already plucked the low-hanging fruit.

As we've seen, accelerating technological progress is not necessarily sufficient for collapsing the distance between our present selves and the higher-hanging fruit, or even of maintaining the pace of progress achieved in the past few decades. With a commitment just to simple progress, we would no longer saddle all ongoing research with the burden bringing us closer to victory against any of the Big Cs. Simple progress allows us both to manage our expectations and preserve public support for the cause.

Here, we can build on a distinction identified by philosopher Tsjalling Swierstra in his own discussion of the harms posed by technological advances. He points out that emerging technologies can have hard impacts that are easily characterized, quantified, and recognized ("poisoning, exploding, polluting and depleting"), but also softer impacts that are "qualitative, ambiguous, and/or indeterminate." Traditional approaches to assigning accountability, such as John Stuart Mill's harm principle, tend to omit this second category, which is consequently "dismissed by technology and policy actors as too fuzzy, or too 'soft,' to take seriously" (Swierstra 2015, p. 7).

Yet it is undeniable that new technologies can have an impact beyond what is strictly quantifiable. Prominent examples in recent years include journalist Nicholas Carr's (2010) thesis that the internet is "rewiring" our brains and making us "shallower," or social scientist Sherry Turkle's (2012) warning that Facebook is ruining the idea of friendship.

In any case, we can apply a similar hard/soft distinction to the benefits of simple progress against cancer. A hard benefit is that some people with cancer will receive better treatments tomorrow than they do today. But one can also conceive of soft benefits arising from the general expectation that things are getting better both for people who have cancer and for those who are at increased risk of developing it. In this way, simple progress can perform a morally empowering function, inspiring ethical conduct on the part of individuals.

As with social distancing and the use of face masks during the COVID-19 pandemic, a society-wide recognition of the cancer risks posed by second-hand smoke has gradually created its own social pressure. The habit has increasingly come to be seen as uncouth and "un-cool," compared to the cultural purchase it once held.

Insofar as red meat, certain pesticides, and other products turn out to be associated with higher risks of cancer, these too are more likely to be phased out by societies that are still "at war" with cancer than in those that are not.

But we should be careful here. While technological progress can be morally empowering, it also can induce moral laziness, particularly when it holds out the promise of silver-bullet solutions. As we've seen, techno-hype can be dangerous. The more that we expect a quick fix for problems like climate change, the less likely we are to do our part to reduce our own carbon footprints. In earlier chapters, we gave the example of CO_2-consuming nanobots that could be released into the atmosphere to keep global warming in check.

Nothing about this proposal is in obvious violation of the laws of physics or logic. And yet, we should know that the horizon bias will make such innovations seem closer than they are. The normal science needed to ensure that such a technology works the way we want it to would be long and torturous.

As such, we should see the carbon-eating-nanobot proposal for what it is: a manifesto science-based siren call that will lull us into a dangerous complacency. Since the engineers are hard at work on the problem, why not reap all of the benefits of cheap fossil fuels while we can? On the contrary, since we're gambling with our children and grandchildren's futures, we ought to approach the ethical implications of technological progress in exactly the opposite way.

The Limits of Progress

Over the past decade, there has been a slew of books, like Pinker's, championing the cause of progress in response to a generalized sense of malaise across the advanced economies. Such titles include, *Progress, The Progress Paradox, Infinite Progress, The Infinite Resource, The Rational Optimist, The Case for Rational Optimism, Utopia for Realists, Mass Flourishing, Abundance, The Improving State of the World, Getting Better, The End of Doom, The Moral Arc, The Big Ratchet, The Great Escape, The Great Surge, The Great Convergence* and *Factfulness: Ten Reasons We're Wrong About the World—and Why Things Are Better Than You Think* (Pinker 2018, p. 52). As HumanProgress.org, a project of the libertarian CATO Institute, notes, (and Marian Tupy earlier stated; Tupy 2013, no page) "historical evidence makes a potent case for optimism. Yet optimism about the current state and future well-being of humankind is difficult to come by" (Lincicome 2020).

Instead, there is a widespread feeling that things are getting worse. Reporting on a YouGov survey in the United Kingdom in early 2015, the *Guardian* noted that fully "71% of people think the world is going to the dogs" (Etchells 2015). Even during the boom years before the 2008 global financial crisis, half of US adults polled by Pew said that their children would be "worse off than people are now" (Taylor et al. 2006).

To the evangelists of progress, these findings depict societies that have lost touch with reality, succumbing to Romanticism, "negativity bias," and nostalgia in the face of mountains of data showing that the "world is getting better." We know that the technologies available to today's children are far more powerful and efficient than anything their parents had. The cheapest smartphones of tomorrow will be able to do things that are impossible for the best iPhones, Samsungs, or Huaweis today. Video streaming will continue to improve, such that the televisual extravaganzas of 2040 will make *Game of Thrones* look quaint (whether the quality of the storytelling is any better, of course, remains to be seen).

But remember: most of these technological improvements will be prone to hedonic normalization. There is no reason to think that the subjective pleasures offered by a Huawei device in 2040 will be markedly better than those offered by an iPhone in the early 2020s. Children born in this decade will have their own technological baseline. There is nothing incompatible in the idea that future generations will have better technology than anyone else in history and yet still be worse off than their parents and grandparents were.

Implicit in this problem is a distinction between moral and material progress. A society makes moral progress when it abolishes slavery, and material progress when it boosts household incomes, provides sustenance, and so forth. One standard metric for material progress is average life expectancy. Since cancer has long been one of the leading causes of death, finding the cure for it has long featured prominently on society's list of material aspirations. Other such goals are to eliminate automobile accidents by replacing human drivers with safe autonomous vehicles; to secure a "Planet B" by colonizing Mars; to "end aging"; and so on.

Sometimes material progress represents a form of moral progress, and sometimes it does not. Some social scientists argue that the ubiquity of smartphones has created an epidemic of mental illness and depression in young people (Twenge 2017). The obesity epidemic across many advanced and developing economies alike would seem to suggest that there is a dark side to progress against hunger.

At the same time, most people agree that modern welfare-state provisions to support the elderly, the disabled, and other marginalized and vulnerable groups constitute a moral victory. Particularly at a time when lifespans are steadily increasing, most Americans would regard a return to the days before Social Security as simply barbaric (though there are plenty of economists who would like to see people continue working further into their silver years).

Here, one can see the soft impacts of material progress. Owing to improvements in material conditions over the decades, our collective commitment to protect vulnerable subsets of the population has strengthened. But that is because many of us will have children and nearly all of us will grow old.

Of course, if today's progressivists are right that we are all in denial about the facts of progress, such widespread ignorance could limit the soft benefits offered by a collective sense of improvement. Yet even if one accepts that there has been steady improvement across many measures of human development and well-being, the problems like cancer and climate change still lend themselves more to a declinist perspective.

Like Sontag back in the 1970s, many people regard these as by-products of modern life, which requires massive energy use and puts us into frequent contact with plastics, toxic fumes, unhealthy foods, and other potential carcinogens. For all of the breakthroughs in cancer prevention, overall cancer incidence (new cases per year per 100,000 people, adjusted for age) was higher in 2016 than it was in 1974, according to the US National Cancer Institute (NCI 2022).

But, again, a glass-half-empty narrative and a glass-half-full narrative need not contradict one another. The progress evangelists are certainly correct that there has never been a better time than now to be diagnosed with a disease like cancer. They are focusing squarely on progress *from* a more primitive past—a period that they urge us not to romanticize. By celebrating the simplicity and supposed moral purity of bygone eras, we imagine away all of their hardships. When we think of the 1870s, we prefer to think in terms of *Little House on the Prairie*, while overlooking small-pox and the lack of educational opportunities for women and girls (Ridley 2010).

For his part, Vincent DeVita believes that nostalgia actively undermines our sense of progress against cancer. "The 30-year time span since the passage of the Cancer Act has distorted our focus," he wrote in 2002.

"Not too many of our current professionals have lived through all the changes. Therefore, there aren't many who can put them into proper perspective. And to each cancer patient today, diagnosis and treatment is still a very difficult experience but they cannot really appreciate how much more difficult it was in 1971" (DeVita 2002, p. 354).

No doubt we do fail to appreciate how far we have come since 1971. But this assessment of progress can be framed another way. We can reflect on what we in 1971 would have accepted as adequate progress by the futuristic-sounding year of 2024. Nixon and his advisers expected that the war would be won within 5 years, and thus surely would be disappointed by today's cancer statistics. Though the current therapies would look like futuristic marvels, they have come nowhere close to providing the cure.

The evangelists of progress ignore this cause for disappointment. Yet it is not enough to be told how much worse it was to get cancer in the 1970s than it is today. What we also need is a realistic sense of what to look forward to, and this is particularly hard to come by. Many of the best-informed among us tend to overestimate the potential of current research. Worse, the story of our fight against cancer has always been Sisyphean: each significant step forward cannot deliver the cure and thus has the effect of pushing the final destination even farther away. The more we have learned about cancer, the more complex it has turned out to be.

It may sound like we are too dour on the prospects for progress. After all, academic research on materials science, management, and even moral development continues to amass at ever-increasing pace. Given that academic research should produce new knowledge, how could it be possible for solutions to material and moral problems to continue to elude us in the months and years to come?

The intense commercialization of academic research appears to have incentivized academic publishers to first restrict access to good research, and then to lower their standards and publish more research that fails to generate any new knowledge. Circa 2012, the Cost of Knowledge campaign focused on the exorbitant prices charged by Elsevier for access to published research (Heyman et al. 2016). In response to the campaign, several editors of high-profile Elsevier journals quit their roles and many academics pledged to forego publishing in Elsevier journals (Heyman et al. 2016). The effect was far from devastating for Elsevier. Like the other major academic publishers, Elsevier pivoted toward a semi-open access publishing model. By making some research free for anyone to read, academic publishers could dodge the accusation that they were making it too difficult for people to access and benefit from research. However, as discussed by Weijers and Jarden (2017), the academic publishers charged academic authors or their funders to make individual research papers free to read while still charging institutions to access the journals that included some of the free-to-read articles. This canny move by academic publishers meant that they created a new revenue stream while undermining the criticism that they were restricting access to valuable research.

This move to author-pays open access publishing is highly concerning because it incentivizes academic publishers to publish every submission, regardless of the value it contributes to our collective knowledge. When combined with academics' personal incentives to publish, author-pays open access publishing will likely result in a deluge of uninformative and distracting research that risks drowning out the research that could enable real material and social advances. As long as profit-motivated companies control academic publishing, we should not expect anything like a quick advance to exponential material or moral progress. Rather, we should view it as a profitable investment opportunity for those lacking the stomach to buy shares in armaments companies just before a major war breaks out.

The problem, then, is not a lack of progress (we are not going backwards, after all). It is rather the persistent failure to satisfy what we perceive as legitimate expectations for that progress. Owing to the promises made by scientists and politicians—all of whom are channeling the modern creed of "infinite perfectibility" (Condorcet 1795, p. 287)—we have a deep-seated sense that cancer should have been cured by now. Many of us live in societies where average incomes have increased by as much as 3000% since 1800—so what's the hold up? (McCloskey 2016).

Because we perceive the progress against cancer to be "too slow," we live in a state of constant frustration, ever susceptible to hype. Each new potential advance against the disease tends to be reported in two ways. First, there are human-interest stories about the lucky few patients who stand to benefit directly from the breakthrough; their own individual cancer may indeed be cured. But then comes the wider speculation, always around the question of whether "this time is different." So often has "the cure" been anticipated that we are left feeling like the villagers in Aesop's fable, "The Boy Who Cried Wolf." It is little wonder that we now meet techno-optimist claims about cancer with a degree of cynicism that simply was not there in the early days of the War on Cancer.

Material Encouragement

Nonetheless, we should recognize that pessimism can have a constricting effect on our willingness to take risks or make sacrifices for the benefit of others. When people think the world is "going to the dogs," they tend to become more defensive and self-seeking, thinking in zero-sum terms and jealously guarding their hard-earned gains against the illicit claims of others.

By contrast, when we collectively sense that we are making material progress, we will all be more willing to pitch in. If we believe that we and everyone else are rising together, we can expect that our own comparatively small sacrifices will be paid back in spades. If we trust that our children have a good chance of being better off than people are today, we will act to make that projected outcome a self-fulfilling prophecy, doing what we can to effect positive change in the world.

Here, we can compare the outlooks of Donald Trump and Lyndon Johnson. In 2019, Trump, representing a cohort of whites who object to the current trajectory of demographic change in America, told migrants and refugees crossing the US-Mexico border, "We can't take you any more. Our country is full" (Guardian Staff 2019, no page). By contrast, when shepherding through the Great Society—which included civil- and voting-rights legislation, Medicare, Medicaid, aid for education, the arts, public broadcasting, urban and rural development, and public services—Johnson could rely on widespread expectations of greater prosperity to come. As the historian Rick Perlstein explains, "Johnson successfully framed his reformist agenda as something that was not ideological at all – conservative, even, simply a pragmatic response to pressing national problems, swept forward on ineluctable tides of material progress" (Perlstein 2009, p. 6).

In other words, the moral advances brought by the Great Society were made possible not merely by economic growth but by the widespread awareness of that growth—by the sense that things were moving in a positive direction. Under these conditions, it made sense to spend money redressing social, economic, and historical injustices, and to invest in an expansion of the "American Dream." This was a time when Americans were planning to go to the moon, and it was just following that feat of technological progress that Nixon decided to take on cancer. Notwithstanding everything we know now about Nixon the man, we can assume that the president and his advisers in 1971 were not acting out of pure cynicism in this case. The same cannot be said for Trump's hollow promise that "We will be ending the AIDS epidemic shortly in America, and curing childhood cancer very shortly" (Porter 2019).

The good news is that we don't actually need much objective progress in order to generate an intersubjective sense of forward movement. Consider the perspective of diabetics, which we touched on in Chapter Four. By and large, diabetics have ample reason to be disappointed about the progress made toward extinguishing the Flame of Hope. We are now approaching the centennial of the discovery of insulin, and diabetes is very much still a thing. If a diabetic in the 2020s could send an email to himself back to the 1990s, that earlier self would be dismayed about the future state of play. From what he had read and been told at the time, a cure was supposed to be just around the corner. But, looking back, we can see that such forecasts were sheer hype, underwritten by the horizon bias.

Nonetheless, it is not as if there has been no progress at all with respect to diabetes. If the insulin-delivery technologies that were used in 1990s were identical to those being used today, diabetics could conclude that there has been no progress. But this isn't the case: today's devices are clearly better. There has been manifest progress, though not of the kind that requires one to adopt faith-based ideas about the inherent direction of history or the uncontested power of humankind over nature. We can be aware of such simple progress without lighting soon-to-be-extinguished flames or building monuments to commemorate truly permanent victories.

Today, a growing number of diabetics and people with various heart conditions are becoming cyborgs, which the Oxford English Dictionary defines as, "A person whose physical tolerances or capabilities are extended beyond normal human limitations by a machine or other external agency that modifies the body's functioning; an integrated man–machine system."

Medical-device cyborgs are not like the homicidal machines in the *Terminator* movies, where the only human part of the synergy is the epidermis. Rather, they include diabetics with smartphone-linked blood-sugar sensors in the arm, and cardiac patients with pacemakers or implantable cardioverter defibrillators. Especially now that these technologies are widely used, they are not considered "game-changers." And yet, they clearly confer both hard and softer benefits for today's patients. So long as they remain affordable and available to those who need them, they will represent simple progress.

These slow and steady improvements show that it doesn't necessarily take much to sustain a sense of progress. It may be sufficient simply to know that improvements of some kind are coming, and will continue to come. We can speculate that Banting in the 1920s would have been deeply disappointed to learn that diabetes still imposes a burden on the society of the 2020s. But that surmise does not prevent today's diabetic cyborgs from looking at their sensors and feeling a sense of progress.

With something like cancer, then, the challenge is to resist hype and disabuse ourselves of hopes for an imminent cure while at the same time maintaining a sense of progress, so as to reap the soft benefits. We need to strike a balance, accepting our limits without ever accepting anything that feels like defeat. Even if we decide to hand off control of the disease to powerful machine learners, it will still be incumbent upon us to manage our collective expectations about what progress can deliver, and when. We will need to keep a clear sense of what we want out of progress against all disease, lest Dr. Robot take us to a destination that is paradise for a few but hell for the rest.

In 1788, Benjamin Franklin speculated that within the next two or three centuries, "if the Art of Physic shall be improv'd in proportion with other Arts, we may then be able to avoid Diseases, and live as long as the Patriarchs in Genesis" (Franklin 1840, p. 348–349). In envisaging millennial lifespans, he was channeling the spirit of the age in which we are still living. Goethe probably would have allowed him the flight of fancy, provided that he did not mistake the horizon for the edge of the earth.

References

Abi-Habib, Maria. 2020. Millions Had Risen Out of Poverty. Coronavirus Is Pulling Them Back. *The New York Times*. 30 April.

Acton, Gemma. 2016. Facebook's Zuckerberg Calls on Citizens to Make the World a Better Place. *CNBC*, 10 November.

Bauman, Zygmunt. 2012. *Liquid Modernity*. Wiley.

Benjamin, Walter (Edited by Hannah Arendt). 1955/2019. Illuminations. First Mariner Books.

Boteach, Melissa, Erik Stegman, Sarah Baron, Tracy Ross, and Katie Wright. 2014. *The War on Poverty: Then and Now*. Center for American Progress. 7 January.

Bury, J.B. 1920/2006. *The Idea of Progress: An Inquiry Into its Origin and Growth*. BiblioBazaar.

Carr, Nicholas. 2010. *The Shallows: What the Internet is Doing to Our Brains*. New York: W.W. Norton & Company, Inc.

Case, Anne, and Angus Deaton. 2020. *Deaths of Despair and the Future of Capitalism*. Princeton University Press.

CATO Institute Policy Report. 2014. *The Launch of HumanProgress.org*. Cato Institute. January/February.

Chernyshevsky, Nikolai (Translation Michael R. Katz). 1863/1989. *What Is to Be Done?* Cornell University Press.

Condorcet, Antoine-Nicholas. 1795/2009. *Outlines of an Historical View of the Progress of the Human Mind*. G. Langer.

Cook, Gareth. 2019. The Economist Who Would Fix the American Dream. *The Atlantic*. July 17.

De Waal, Alex. 2021. *New Pandemics, Old Politics*. Polity.

Desilver, Drew. 2018. *For Most U.S. Workers, Real Wages Have Barely Budged in Decades*. Pew Research Center. 7 August.

DeVita, Vincent T. 2002. A Perspective on the War on Cancer. *The Cancer Journal*, September–October 8 (5): 354. https://doi.org/10.1097/00130404-200209000-00002.

Eckermann, Johann Peter. 1930/1998. *Conversations of Goethe*. Da Capo Press.

Elias, Amanuel, and Yin Paradies. 2021. The Costs of Institutional Racism and its Ethical Implications for Healthcare. *Journal of Bioethical Inquiry* 18: 45–58.

Etchells, Pete. 2015. Declinism: Is the World Actually Getting Worse? *The Guardian*. January 16.

Fox-Skelly, Jasmin. 2017. There Are Diseases Hidden in the Ice, and They Are Waking Up. *BBC*. 4 May.

Franklin, Benjamin. 1840. *Letter to John Lathrop*, May 31, 1788. The Works of Benjamin Franklin, Volume 10. Childs and Peterson.

Gott, Richard J. 1993. Implications of the Copernican Principle for Our Future Prospects. *Nature* 363: 315–319. 27 May.

Graeber, Charles. 2018. *The Breakthrough: Immunotherapy and the Race to Cure Cancer*. Twelve.

Guardian Staff and Agencies. 2019. "Our Country Is Full": Trump Claims Emergency During Border Visit. *The Guardian*. 5 April.

Hauser, David J., and Norbert Schwarz. 2019. The War on Prevention II: Battle Metaphors Undermine Cancer Treatment and Prevention and Do Not Increase Vigilance. *Health Communication* 35 (13): 1698–1704. 9 September. https://doi.org/10.1080/10410236.2019.1663465.

Herben, Pierre. 2014. *Dreaming of a Green Desert*. Thomson Reuters Foundation. 5 February.

Hertz, Noreena. 2021. *The Lonely Century: How to Restore Human Connection in a World That's Pulling Apart*. Currency.

Heyman, Tom, Pieter Moors, and Gert Storms. 2016. On the cost of knowledge: Evaluating the boycott against Elsevier. *Frontiers in Research Metrics and Analytics* 1: 7. https://doi.org/10.3389/frma.2016.00007.

Lincicome, Scott. 2020. Some Reasons for Optimism Regarding the "Hollowing Out" of America's Middle Class. *CATO Institute*, 20 August 2020. Retrieved 9 September 2024 from: https://www.cato.org/blog/some-reasonsoptimism-regarding-hollowing-out-americas-middle-class

McCloskey, Deirdre N. 2016. The Formula for a Richer World? Equality, Liberty, Justice. *The New York Times*. 2 September.

Miller, Anthony B., Claus Wall, Cornelia J. Baines, Ping Sun, Teresa To, and Steven A. Narod. 2014. Twenty Five Year Follow-Up for Breast Cancer Incidence and Mortality of the Canadian National Breast Screening Study: Randomised Screening Trial. *British Medical Journal.*, 11 February 348: g366. https://doi.org/10.1136/bmj.g366.

Moss, Howard. 1962. *The Magic Lantern of Marcel Proust*. The MacMillan Company.

National Cancer Institute (NCI). 2022. Incidence. In *Cancer Trends Progress Report*. October 2022.   https://progressreport.cancer.gov/diagnosis/incidence#field_most_recent_estimates. Accessed 21 Nov 2022

Nisbet, Robert. 1980/2017. History of the Idea of Progress. Routledge Taylor & Francis Group New York.

Pardoll, Drew. 2016. How Immunotherapy Helps Cancer Patient's Own Body Attack Tumors. An excerpt from "Chasing Cancer," *Washington Post Live*, December 6. *The Washington Post*, 9 December.

Perlstein, Rick. 2009. *Nixonland: The Rise of a President and the Fracturing of America*. Scribner.

Pickert, Kate. 2019. *Radical: The Science, Culture, and History of Breast Cancer in America*. New York: Little, Brown Spark.

Pinker, Steven. 2018. *Enlightenment Now: The Case for Reason, Science, Humanism, and Progess*. Viking.

Porter, Tom. 2019. Trump Promises Cincinnati Rally that US Will End AIDS Epidemic and Find a Cure for Childhood Cancer "Very Shortly.". *Business Insider*. August 2.

Ridley, Matt. 2010. *The Rational Optimist: How Prosperity Evolves*. Harper.

Rigaud, Kanta Kumari, Alex de Sherbinin, Bryan Jones, Jonas Bergmann, Viviane Clement, Kayly Ober, Jacob Schewe, Susana Adamo, Brent McCusker, Silke Heuser, and Amelia Midgley. 2018. *Groundswell: Preparing for Internal Climate Migration*. The World Bank.

Romm, Joe. 2018. Ocasio-Cortez Says We Need World War II-Scale Action on Climate. Here's What That Looks Like. *Think Progress*. 12 February.

Scheidel, Walter. 2018. *The Great Leveler: Violence and the History of Inequality from the Stone Age to the Twenty-First Century*. Princeton University Press.

Sheffield, Rachel and Rector, Richard. 2014. The War on Poverty After 50 Years. *The Heritage Foundation*. 15 September. Retrieved 4 November 2023 from: http://report.heritage.org/bg2955

Sontag, Susan. 1978. Illness as Metaphor. In *Sontag Essays of the 1960s & 70s*. Library of America.

Swierstra, Tsjalling. 2015. Identifying the Normative Challenges Posed by Technology's "Soft" Impacts. *Etikk i praksis, the Nordic Journal of Applied Ethics* 9 (1): 5–20. https://doi.org/10.5324/eip.v9i1.1838.

Taylor, Paul, Cary Funk, and Peyton Craighill. 2006. *Once Again, the Future Ain't What It Used to Be*. Pew Research Center. 2 May.

Thigpen, Cary Lynne, and Cary Funk. 2019. *Most Americans Say Science Has Brought Benefits to Society and Expect More to Come*. Pew Research Center. 27 August.

Tupy, Marian. 2013. *Human Progress: Not inevitable, Uneven, and Indisputable*. Reason magazine, 30 October 2013. Retrieved 4 November 2023 from: https://reason.com/2013/10/30/human-progress-not-inevitable-uneven-and/

Turgot, Anne Robert Jacques (Edited by David Gordon). 1750/2011. A Philosophical View of the Successive Advances of the Human Mind. *The Turgot Collection*. Ludwig von Mises Institute.

Turkle, Sherry. 2012. The Flight from Conversation. *The New York Times*. 21 April.

Twenge, Jean M. 2017. Have Smartphones Destroyed a Generation? *The Atlantic*. September.

UNHRC. 2018. *Report of the Independent International Fact-Finding Mission on Myanmar*. United Nations Human Rights Council. 12 September. https://www.ohchr.org/Documents/HRBodies/HRCouncil/FFM-Myanmar/A_HRC_39_64.pdf.

Warofka, Alex. 2018. An Independent Assessment of the Human Rights Impact of Facebook in Myanmar. *Facebook*. 5 November. https://about.fb.com/news/2018/11/myanmar-hria/. Accessed 8 Aug 2020.

WarOnCancer.com 2020. https://www.waroncancer.com. Accessed 14 Aug 2020.

Weber, Elke U. 2006. Experience-Based and Description-Based Perceptions of Long-Term Risk: Why Global Warming Does Not Scare Us (Yet). *Climatic Change* 70: 103–120. https://doi.org/10.1007/s10584-006-9060-3.

———. 2010. What shapes perceptions of climate change? *Wiley Interdisciplinary Reviews: Climate Change* 1 (3): 332–342.

Weijers, Dan, and Aaron Jarden. 2017. The International Journal of Wellbeing: An Open Access Success Story. In *Open: The Philosophy and Practices that are Revolutionizing Education and Science*, ed. Rajiv S. Jhangiani and Robert Biswas-Diener, 181–194. London: Ubiquity Press.

Workman, Paul. 2019. Winning the War on Cancer: Why Words Matter. *The Institute for Cancer Research.*. 6 February.